T0177725

The Tools of Metaphysics and the Metaphysics of Science

Theodore Sider is Andrew W. Mellon Chair in Philosophy at Rutgers University. He completed his PhD at the University of Massachusetts at Amherst, and has held positions at Cornell University, New York University, Syracuse University, and the University of Rochester. He is the author of *Writing the Book of the World* (Oxford 2011) and *Four-Dimensionalism: An Ontology of Persistence and Time* (Oxford 2001), and two textbooks: *Logic for Philosophy* (Oxford 2010) and *Riddles of Existence: A Guided Tour of Metaphysics* (Oxford 2005, with Earl Conee).

The Tools of Metaphysics and the Metaphysics of Science

Theodore Sider

CLARENDON PRESS · OXFORD

OXFORD
UNIVERSITY PRESS

Great Clarendon Street, Oxford OX2 6DP,
United Kingdom

Oxford University Press is a department of the University of Oxford.
It furthers the University's objective of excellence in research, scholarship,
and education by publishing worldwide. Oxford is a registered trade mark of
Oxford University Press in the UK and in certain other countries

First published 2020

First published in paperback in 2022

Published in the United States of America by Oxford University Press
198 Madison Avenue, New York, NY 10016, United States of America

British Library Cataloguing in Publication Data
Data available

Library of Congress Cataloging in Publication Data
Data available

ISBN 978–0–19–881156–5 (Hbk.)
ISBN 978–0–19–286476–5 (Pbk.)

DOI: 10.1093/oso/9780198811565.001.0001

For Jane and Polly

For Jane and Polly

Contents

Preface

This book is about issues at the intersection of metaphysics and the philosophy of science, especially the philosophy of physics. It is written in the belief that each of these fields can learn from the other.

Projects straddling fields face an inherent danger: that their forays into the field further from the writer's own will be superficial and engage inadequately with that field's internal concerns. Philosophers of science may find some of my focus alien, overly metaphysical. I am sensitive to this danger, and offer my contributions in a spirit of collaboration.

But philosophers of science sometimes overestimate the gulf between themselves and metaphysicians. They regard metaphysicians as a credulous lot who uncritically assume the intelligibility of questions beyond those justifiable from a sober scientific outlook. Sometimes this is indeed true. But sometimes something else is going on. Philosophers of science often take implicit stands themselves on various metaphysical issues, sometimes without noticing it. Metaphysicians didn't invent metaphysical issues; they simply made them explicit. When trying to investigate what physical and other scientific theories tell us about the nature of reality, it's inevitable that one would bump up against the very general questions about reality with which metaphysicians wrestle.

Thus I hope that philosophers of science will take seriously the issues I raise, and come to see that some of my concerns bear on their own. Concerns from the philosophy of science have certainly influenced my own thinking about metaphysics.

Chapters 2–5 are on, respectively, the relation between properties and the laws of nature, individuals and identity, quantitative properties, and theoretical equivalence. They can mostly be read independently, although Chapter 1, which introduces the conceptual framework of the book, should be read, or at least skimmed, first. Of Chapters 2–5, Chapter 2 (properties and laws) is the most purely metaphysical; philosophers of physics may wish to move quickly to Chapters 3–5 (although section 2.3 introduces an idea that will be important later). The final chapter is a brief synoptic conclusion.

An early draft of this book was the basis for my 2016 Locke Lectures at Oxford University. I am grateful to Oxford University for inviting me to give the lectures, and to All Souls College for hosting me as a Visiting Fellow during my visit. Finally, I am grateful to many friends for help with this project: Frank Arntzenius, David Baker, Elizabeth Barnes, Nathaniel Baron-Schmitt, Karen Bennett, Selim Berker, Alexander Bird, Phillip Bricker, Ross Cameron, Fabrice Correia, Troy Cross, Cian Dorr, Tom Donaldson, Jamie Dreier, Vera Flocke,

Verónica Gómez, Jeremy Goodman, Hilary Greaves, Chris Hauser, Katherine Hawley, Mike Hicks, Nick Huggett, Alex Kaiserman, David Kovacs, Ofra Magidor, Niels Martens, Vivek Mathew, Michaela McSweeney, Elizabeth Miller, Sarah Moss, Daniel Murphy, Jill North, Asya Passinsky, Laurie Paul, Zee Perry, Lewis Powell, Alex Roberts, Gideon Rosen, Ezra Rubenstein, Jeff Russell, Simon Saunders, David Schroeren, Erica Shumener, Jack Spencer, Jason Turner, Gabriel Uzquiano, Mahmood Vahidnia, Isaac Wilhelm, Tim Williamson, and many others. I am grateful to Peter Momtchiloff at Oxford University Press for his support. And I am especially grateful to Eddy Chen, Shamik Dasgupta, John Hawthorne, Nick Huggett, Kris McDaniel, and Jonathan Schaffer for helpful feedback on the entire manuscript.

Theodore Sider
New Brunswick, NJ
29 October 2019

1

Postmodal Metaphysics and Structuralism

1.1 Tools in metaphysics

By "tools in metaphysics" I mean the core concepts used to articulate metaphysical problems and structure metaphysical discourse. They are a lens through which we view metaphysics.

The metaphysical tools of choice change over time, and as they do, the problems of metaphysics are transformed. We view the very same problems through different lenses. In the 1950s and 1960s the preferred tools were concepts of meaning and analysis. So when personal identity over time was discussed, for example, the question was, what are we *saying* when we re-identify persons over time?[1] In the 1970s through to the 1990s, the tools became modal, and the questions of personal identity underwent a corresponding transformation: what conditions governing personal identity hold of metaphysical necessity? Would it be possible to survive the loss of all of one's memories?

The mind–body problem had a similar arc. In the 1950s the goal was to give an analysis of mental concepts, but later the questions became modal; whether, for instance, it would be possible for a world physically like ours to lack consciousness.

Like all philosophical questions, metaphysical questions begin life in vague, primordial form. The mind and the body: what's up with that? How are they related? Before real progress can be made, the questions must be made precise, and placed in a developed theoretical setting. This is the job of tools of metaphysics. With particular tools in hand, the primordial questions begin to seem, in retrospect, as first attempts to ask what was the proper question all along. The proper questions will be viewed as *better* than the questions yielded by rival tools— clearer perhaps, or more precise, substantive, or objective; or better in lacking false presuppositions, or being less susceptible to confusion caused by misleading natural language, or having a better associated methodology, or being more likely to connect with questions outside metaphysics.

---

[1] See Strawson (1959), for instance.

1.2 Postmodal metaphysics

Recently there has been a shift to new tools (or perhaps a return to old ones), which I will call "postmodal". David Lewis (who had also been a leader in the modal revolution) enriched his conceptual toolkit with the concept of *natural properties and relations*—those elite properties and relations that determine objective similarities, occur in the fundamental laws, and whose distribution fixes everything else. I myself have argued for the centrality of a concept that is closely related to Lewis's notion of naturalness: the concept of structure, or as I'll put it here, the concept of a *fundamental concept*. Fundamental concepts are not limited to those expressed by predicates; we may ask, for instance, whether quantifiers or modal operators express fundamental concepts—whether they help to capture the world's fundamental structure. Kit Fine (re-)introduced the concept of *essence*, and argued that it should not be understood modally. He pointed out that although it does seem to be an essential feature of the singleton set {Socrates} that it contain Socrates, it does not seem to be an essential feature of Socrates that he be contained in {Socrates}; being a member of this set is not "part of what Socrates *is*". Thus we cannot define a thing's essential features, as it had been common to do in the halcyon days of the modal era, as those features that the thing possesses necessarily, for it is plausible that Socrates possesses the feature of being a member of {Socrates} necessarily.[2] Fine also (re-)introduced a notion of *ground*. One fact grounds another, he said, if the second holds in virtue of the first—if the first explains, in a distinctively metaphysical way, the second. Interest in ground and related concepts over the past ten years or so has been intense.

Friends of the postmodal revolution think that modal conceptual tools need to be supplemented, or perhaps even replaced, by one or more of these postmodal concepts.[3] A recurring refrain has been that modal concepts are too crude for many purposes, in that even after modal questions are settled, there remain important questions that can be raised only by using the postmodal tools. Fine's example of {Socrates} illustrates this, as does the often-cited example of the Euthyphro question: even after it is settled that something can be pious if and only if the gods love it, there is a further question, that of whether something is pious because the gods love it, or whether the reverse is somehow true. This appears to be a question of ground.[4] Another refrain has been that modal truths are often epiphenomenal, a mere reflection of deeper postmodal structure.

The story of a linear progression from conceptual analysis to modality to fundamentality/essence/ground is an oversimplification. For instance, inspired by Quine's 'On What there Is', much metaphysical inquiry has centred on ontological questions, questions structured by the concepts of ontology (for Quineans, first-order existential and universal quantification). From 1980–90, three of the major

---

[2] See also Dunn (1990, section 4).

[3] See Bennett (2017); Fine (1994a; b; 2001; 2012); Rosen (2010); Schaffer (2009); Sider (2011).

[4] See Evans (2012).

works of metaphysics were focused on ontology: Field's *Science without Numbers*, Lewis's *On the Plurality of Worlds*, and van Inwagen's *Material Beings*.[5]

Nevertheless, the final transition in the simplified story is what will be important here: the shift from modal to postmodal tools. I'm interested in how the shift affects first-order metaphysical questions. (I'm also interested in the reverse direction of influence, what the tools' repercussions for first-order questions can teach us about the tools. As we will see, in certain contexts, particularly in the metaphysics of physics, the appropriate tool is fundamentality, rather than essence or ground.) The postmodal revolution has been very "meta", about what we're asking when we ask metaphysical questions. But the choice of tools also affects the questions' answers. The matter of tools isn't purely methodological, or more a priori, or anything like that. It isn't "first metaphysics", in the sense that it must be done before, and in isolation from, the rest. It's just more metaphysics, albeit especially intertwined with a wide range of other questions.

1.3 Structuralism

If this book has a single thesis, it is that the choice of metaphysical tools matters to first-order metaphysics, especially when it comes to "structuralist" positions in the metaphysics of science and mathematics.

'Structuralism' is pretty vague, but the idea is that patterns or structure are primary, and the entities or nodes in the pattern are secondary.

The argument for structuralism is often epistemic: our evidence is only for patterns. One could respond with a merely epistemic doctrine: all we know is the pattern; what instantiates the pattern is real but unknown.[6] But structuralists respond metaphysically: the patterns are metaphysically, not just epistemically, primary.

Such epistemic arguments have close nonepistemic cousins: that mere differences in nodes are distinctions without a difference. And there can also be entirely nonepistemic arguments, such as that dispensing with the nodes while keeping the structure yields a simpler picture of the world.

Structuralist positions have been defended in a number of different areas in the metaphysics of science and mathematics (and elsewhere). I will focus on three: nomic essentialism, comparativism about quantity, and structuralism about individuals.

According to nomic essentialism, networks of nomic, or lawlike, relations between properties are primary and the properties themselves are secondary. When a law of nature governs a property, this isn't something that just happens

---

[5] And indeed, Schaffer's (2009) defence of ground focuses on the limitations of a purely ontological approach to metaphysics more than on the limitations of a purely modal approach.

[6] Examples include what Ladyman (1998) calls epistemic structural realism and the "humility" theses of Langton's (1998) Kant and Lewis (2009).

to the property. The nature of the property itself is somehow bound up with the laws governing it and other properties.

Why believe such a claim? One putative reason is epistemic. What we know of the property of charge (for example), we know through its nomic profile: entities with this property are correlated, by law, with the electromagnetic field, which is in turn correlated with the motions of other particles, depending, in part, on their charges. What do we know of the property of charge *in itself*? Nothing—we know of it only as "that which is correlated, by law, with such-and-such". So why assume that there *is* anything more to the property than this lawful correlation?

That was nomic essentialism, but there are also the closely related doctrines of dispositional and causal essentialism, according to which, respectively, the dispositional and causal roles of properties are prior to the properties themselves.

Another form of structuralism pertaining to properties concerns quantitative properties, those that can be measured by numbers. Charge and mass, for instance, come in degrees, which we represent with numbers. Now, for any distribution of values for a given quantity across all individuals—an assignment of 2 g mass to this thing, of 1 g mass to that thing, and so on—there is a network of corresponding relations amongst those individuals: one individual is twice as massive as another, a certain pair of individuals are together exactly as massive as a certain other pair, and so on. According to a "comparativist" view of a given quantity, the network of relations is prior to the individual values for that quantity. Like nomic essentialism, this form of structuralism can be supported on epistemic grounds: we observe relational rather than absolute quantitative facts, as when we use a set of scales to establish that two things are exactly as massive as each other.

Yet another form of structuralism pertains to individuals: the network of qualities—properties and/or relations—had by individuals is primary and the individuals themselves are secondary. And again there is an epistemic argument. We seem to have no way to distinguish between the following two arrangements:

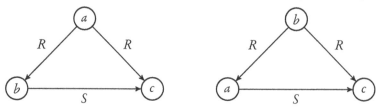

Observation tells us only the qualities of individuals, and not which individuals they are; individuals don't have metaphysical nametags. So why suppose that there exists something beyond the qualities, an extra fact of which things occupy which places in the network of properties and relations, which can vary independently of the network? Why suppose there's a different possible world that is qualitatively just like ours, except that Barack Obama and I have "exchanged places", so that I am a 6-feet-1-inch-tall politician born in a state known for its beaches and volcanoes, and he is a 5-feet-9-inch-tall philosopher from a city known for its cheesesteaks and unruly sports fans? Why not suppose instead that the identities of individuals cannot vary independently of the pattern, and indeed that the pattern is all there is?:

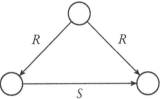

Structuralisms about individuals have been defended within pure metaphysics—the bundle theory, for example. And recently such a position has been developed in the philosophy of science: structural realism. A related position is also defended in the philosophy of mathematics, only here the chief argument is not that all we *observe* is the pattern—since we don't observe mathematical entities—but rather that the pattern is all that matters to the practice of mathematics. What's distinctive of the natural numbers, for instance, is that they be an 'infinite series each of whose members has only finitely many precursors', as Quine (1960b, p. 242) put it. It doesn't matter to mathematics which individuals are in this structure or what their intrinsic features are; all that matters is that they be so structured. So perhaps all there is to the natural numbers is this structure.

1.4 Modal and postmodal structuralism

All this talk of patterns being "primary", of patterns being "all there is", is extremely vague, and how it is precisified depends on the metaphysical tools one adopts. For instance, using modal concepts one can say that nodes and patterns cannot vary independently; and many structuralist positions have in fact been formulated in this way.

One form a modal structuralist thesis can take is this: the pattern cannot vary while the nodes remain constant. Dispositional essentialism, for instance, has usually been articulated as the claim that the very same properties and relations could not have existed while having different dispositional features; the network of dispositional relationships amongst properties and relations cannot vary while the identities of those properties and relations remain constant. Modal structuralist theses can also take the converse form: nodes cannot vary while the pattern remains constant. Structuralism about quantities can be understood as the claim that any two possible worlds that are alike in their distribution of quantitative relations (relations like being-twice-as-massive-as) are alike simpliciter with respect to quantities; thus doubling everything in mass does not result in a different possible world. Structuralism about individuals can be articulated as antihaecceitism, the claim that it's impossible for individuals to vary independently of qualitative facts—that is, that there are no two possible worlds that have the same distribution of qualities over individuals, but in which different individuals occupy different qualitative roles; there is no duplicate possible world in which I have exchanged places with Barack Obama.

Now, in the case of mathematical individuals, no one construes structuralism modally, because facts about mathematical entities are generally taken to be necessary. Anyone who accepts this dogma already thinks that it's necessarily true that the number 1 occupies its place in the structure of natural numbers, for instance, and yet it's often thought that some question of structuralism remains open.[7]

From a postmodal point of view, the failure of modal tools to articulate a meaningful thesis of mathematical structuralism is a sign of a deeper problem. *Any* modal thesis is bound to be unsatisfying as a formulation of any form of structuralism, because modality is "insensitive to source", as Fine (1994a, p. 9) puts it. A modal structuralist thesis says that independent variation of patterns and nodes is impossible, but says nothing about *why* this is impossible; the impossibility might be due to something that, intuitively, has nothing to do with structuralism. This is made vivid by an example due to Shamik Dasgupta (2011, p. 118). Suppose that a very surprising "Spinozistic" thesis is in fact true of modal reality, namely that all truths are true necessarily. Then each modal structuralist thesis would automatically be true. Nodes and patterns can't vary independently because nothing can vary at all. But this would not be because of any priority of patterns over nodes; it would be because of the quirky nature of modality.[8] A more satisfying statement of a structuralist position will no doubt *imply* a modal thesis, but that modal thesis would be due to some deeper nonmodal thesis: nodes and patterns can't vary independently because nodes and patterns are tied together in some nonmodal way. For example, a postmodalist won't take antihaecceitism as the statement of a structuralist position since antihaecceitism is a modal thesis, but will seek instead some nonmodal formulation, for example the thesis that individuals just are bundles of universals. This thesis *implies* the modal thesis (given plausible principles connecting modality to claims of the form 'X just is Y'[9]), but is a distinctively structuralist claim about the nonmodal tie between individuals and qualities.

Postmodalists have a similar attitude to modal formulations of many other metaphysical doctrines, not just structuralism. The modal thesis of mind-body

---

[7] The problem could be avoided by denying that mathematics is necessary. But if this denial is because of a more general claim that the necessary truths are "minimal" (see later in this section), that minimality claim (if not underwritten by some postmodal thesis) might be inconsistent with the modal articulation of mathematical structuralism.

[8] See also Fine's (1995, p. 271) point about the Tractarian view that all objects exist necessarily. As John Hawthorne pointed out to me, these arguments are perhaps less decisive than they initially appear. A friend of modality might insist that Spinozism *would* obliterate questions of structuralism, much as a friend of ground would need to insist that bizarre theories of ground, such as that nothing grounds anything, would obliterate questions of structuralism. The latter insistence strikes me as more reasonable, but others may disagree.

[9] See Dorr (2016) and Rayo (2013) on this sort of language.

materialism, that there is no mental difference without a physical difference, is all well and good, but to what is it due? What is it about the nature of mind that rules out the possibility of independent variation of the mental? A satisfying materialism would give some answer, such as that there are no fundamental mental properties or relations, and that all fundamental properties and relations are physical.[10]

I'm going to assume that modal articulations of the structuralist positions to be considered in this book are indeed inadequate, and further, that postmodal articulations are needed. Though I won't say much in support of this assumption, it's worth distinguishing some different groups of philosophers who would accept it.

One group would oppose modal articulations of structuralism because they think that modality is *nonfundamental*. If modality is nonfundamental then any modally-articulated structuralist thesis would not itself be fundamentally true, but would rather be due to certain facts about fundamental reality; and a structuralist might prefer to articulate those facts directly. To be sure, it isn't true in general that metaphysical theses must always be articulated in perfectly fundamental terms. Criteria of persistence for entities falling under nonfundamental sorts (say, persons) are most appropriately stated using nonfundamental concepts (psychological concepts, perhaps). To take another example, the causal structures at issue in various branches of social metaphysics emerge only in terms of higher-level concepts (Barnes, 2014). But according to this first group, the structuralist theses we are discussing are different: unlike theses of higher-level persistence or higher-level causal structure, they are meant to be theses about fundamental reality, and ought to be concisely stateable in fundamental terms.

A second group would oppose modal articulations of structuralism because they think that modality is not only nonfundamental, but also metaphysically *superficial*. On my own view, for instance, the necessary truths are just certain truths that we "hold constant" when talking about alternatives to actuality, and the distinction between truths we hold constant in this way and truths that we don't hold constant is more or less conventional.[11] Given this approach, if a structuralist thesis aspires to articulate something metaphysically important, it should not do so via the metaphysically superficial language of modality. At best this would be a misleading way to get at an important nonmodal fact, and at worst it would not reflect anything important at all.[12]

A third group thinks that although necessity is metaphysically deep, and perhaps even fundamental, the necessary truths are *minimal*. (Equivalently, they think that the possible truths are plentiful.) Suppose, for example, you think that,

---

[10] It's tempting to regard many of the contortions philosophers of mind underwent to construct a proper modal formulation of materialism as the result of struggling to find a modal proxy for a simple idea about what is fundamental.

[11] See Sider (2011, chapter 12). Sidelle (1989) holds a similar view; see also Nolan (2011).

[12] To be sure, a shift to postmodal concepts wouldn't improve the situation if those concepts were themselves superficial; see Dasgupta (2018).

with a few exceptions (logical truths, perhaps), no truth is necessary unless it is underwritten by some postmodal claim (such as that individuals just are bundles of universals). You will then be dissatisfied with modal articulations of structuralism, for you will think they can't be true unless underwritten by some appropriate postmodal claim.

Finally, a fourth group thinks that modal structuralist theses may well be true, metaphysically deep, and even fundamental, but nevertheless are *unsuitable* statements of structuralism because they are not supported by structuralist arguments. Consider the argument that permutations of individuals amongst qualitative roles are distinctions without a difference. The modal formulation of structuralism about individuals—antihaecceitism—wouldn't, in my view, be supported by this argument, since it doesn't imply that permutationally different scenarios aren't different; it just implies that they aren't both possible. Or consider epistemic arguments that only structuralism can explain our knowledge of the domain in question; one might think that a merely modal formulation of structuralism, even if true, couldn't explain our knowledge.[13]

1.5 The challenge for postmodal structuralism

The demand for postmodal formulations of metaphysical doctrines can make a difference: there is no guarantee that a given doctrine *can* be formulated post-modally.

One obstacle is that there may not be any coherent postmodal thesis in the vicinity. Consider structuralism about individuals, a view which seeks to somehow "downgrade" particular individuals relative to their qualitative structure. The most straightforward kind of downgrading is elimination: fundamentally speaking there exist no individuals, only a structure. But this appears to make no more sense than the Cheshire Cat's lingering smile. For what a qualitative structure *is*, is some individuals instantiating properties and relations.

I don't mean to suggest that no response is possible—hence the term 'obstacle'. There are other ways one might attempt to downgrade the metaphysical status of individuals, as we will see in Chapter 3. The question is whether any account is both coherent and avoids other obstacles.

(Many structuralist views merely prioritize relations over properties, rather than prioritizing relations and properties over the entities that instantiate them, and hence don't face this obstacle. For example, meaning holists claim that meaning ultimately consists, not in the possession of semantic properties by individual words or sentences, but rather in a network of semantic relations across all words or sentences. This view makes a claim about which kinds of features are present in the most fundamental semantic facts—relations, not properties—but the existence of the entities possessing those features—words, or sentences, understood in some

---

[13] Compare Dasgupta (2011).

nonsemantic sense—isn't denied or understood structurally. This all is perfectly straightforward, metaphysically.)

Another obstacle is that there might be a conflict with "postmodal logic". A natural strategy for formulating structuralism appeals to ground: facts about the pattern somehow ground facts about the nodes. And it's natural to take "facts about the pattern" to be existentially quantified facts whose instances are facts about nodes. Thus existential facts would ground their instances. But the usual logic of ground demands the reverse: instances ground existentials. The problem, again, simply doesn't arise if one articulates structuralism in merely modal terms. Ground is a hierarchical notion—facts are arranged in a hierarchy of more or less basic facts according to certain rules—and this additional imposed constraint can conflict with a structuralist thesis.

A third potential obstacle is that even if a modal position can be "translated" into a coherent and consistent postmodal thesis, that thesis might be theoretically unattractive from a distinctively postmodal point of view. For instance, if a postmodal structuralist thesis is a claim that certain concepts are fundamental, it may be that the concepts required to state the structuralist thesis are complex in certain objectionable ways, or cannot be used to state suitable laws of nature—complaints that flow from a natural epistemology for fundamentality, as we will see.

The preceding was not intended as a blanket argument against all forms of structuralism. 'Structuralism' is too broad a term to allow for meaningful debate at such a level of generality; we must examine each case individually. Still, I do think that in some cases, structuralism is an idea that looks good when viewed through the metaphysically superficial lens of modality, but becomes much less attractive when we turn up the metaphysical resolution.

We will discuss all this—the various forms of structuralism, and the concerns about what they might amount to in postmodal terms—in more detail in subsequent chapters. For the remainder of this chapter let us look more closely at various postmodal concepts, beginning with essence.

1.6 Essence

Fine's example of Socrates and the singleton set of Socrates is the intuitive heart and soul of the contemporary discussion of essence: it is meant to convince us that there is a real distinction between those facts or features that are, and those that are not, part of a given thing's nature; and it is thought that this distinction cannot be captured in modal terms.

Fine explores various ways to formalize discourse about essence; we can focus on the regimentation $\Box_{x_1, x_2, \ldots} A$, which says that A holds in virtue of the natures of entities $x_1, x_2, \ldots$. Thus the true claim that it's of the essence of {Socrates} to have Socrates as a member would be regimented as:

$$\Box_{\{Socrates\}} \text{ Socrates} \in \{Socrates\}.$$

and the false claim that it's of the essence of Socrates to be a member of {Socrates} would be regimented as:

$$\Box_{\text{Socrates}}\ \text{Socrates} \in \{\text{Socrates}\}.$$

As we've seen, Fine denies that essence should be defined in terms of necessity. Indeed, Fine accepts the reverse direction of definition: a necessary truth is a truth that holds in virtue of the essences of all things.[14]

1.7 Ground

Turning next to ground, we may again begin with Fine's regimentation: one or more facts $f_1, f_2, \ldots$ are said to ground another fact, g:

$$f_1, f_2, \ldots \Rightarrow g.$$

There are many subtle details which I'll mostly ignore or elide: I'll alternate between speaking of grounding of facts, propositions, and speaking of grounding using a sentence operator, and I'll mostly (though not always) ignore distinctions between full, partial, strict, weak ground, and the like.

Philosophers often speak of facts "holding in virtue of", "being grounded in", "depending on", "consisting in", "being explained by", or "being made true by" other facts. As Gideon Rosen (2010) vividly recounts, we have long viewed such talk with suspicion, preferring instead allegedly clear modal and other language, at least when we're trying to be rigorous. But Rosen, Fine (2001; 2012), Jonathan Schaffer (2009), and many others now say that such talk is legitimate after all. It concerns a relation of grounding, which is an irreplaceable conceptual tool in philosophy.

Claims of grounding presumably *imply* modal claims: if f grounds g then f necessitates g.[15] But the converse implication doesn't hold: even if it happens to be necessarily true that g obtains whenever f does, there may not be the right sort of connection between f and g so that f grounds g.

---

[14] Fine (1994a, p. 9). Incidentally, I doubt this is right. There are some subject matters where the truth is necessary, whatever that truth happens to be, but where the truth isn't settled by the essences of the entities involved. For example, for a certain sort of realist about set theory, either the continuum hypothesis or its negation is true; and whichever is true is necessarily true. But this doesn't seem to be settled by essences (by the essence of set-membership, say). It's just a fact about what sets happen to exist. (See Sider (2011, p. 267).) Similarly for the principle of universal composition, according to which any plurality has a mereological sum: it's necessary if it's true, but its truth doesn't seem to be due to essences. Fine himself might bring in his postulational account of existence (2007) to claim that these truths are essential after all: the idea might be, in the case of set theory, that we can choose which notion of set-membership to adopt, and that sets obeying the laws corresponding to that notion are thereby postulationally introduced, with truths about them holding in virtue of the essence of the chosen notion. But this reply seems unavailable given a more orthodox Platonist conception of mathematical existence.

[15] Although this is so according to "grounding orthodoxy" (e.g./i.e. Fine (2012)), others uphold some weaker connection to modality. See Bliss and Trogdon (2016, section 5).

Many of the traditional questions of philosophy, it is said, are really about grounding. The question of moral naturalism, for instance, should really be understood as the question of whether moral facts are grounded in natural facts. It is a distortion to understand the question in modal terms, for instance as the question of whether moral facts are necessitated by natural facts, since according to many moral nonnaturalists, even though moral facts are "above and beyond" the natural facts, they nevertheless cannot vary independently of the natural facts. (And recall Dasgupta's point about the Spinozistic view that every truth is necessary.)

1.7.1 *Ground and levels*

There is a familiar "levels" picture of reality, in which facts at "higher" levels rest on facts at "lower" levels, with everything ultimately based on a ground floor of fundamental facts. Perhaps psychological facts are higher than chemical facts, which in turn are higher than physical facts, which are in turn fundamental facts.

The levels picture has always faced the question of how the levels are related. In his classic discussion of theoretical reduction, Ernest Nagel (1961, p. 354) himself mentioned three views of the status of his "coordinating definitions", which connect higher- and lower-level concepts in theoretical reductions. Coordinating definitions can be analytic, Nagel said, they can be stipulated by fiat, or they can be "factual" or "material". None of these three ideas seems correct as an account of the relationship between facts or properties at different levels. The third idea is apparently that of a relationship of lawful co-variation between metaphysically separate, metaphysically coequal partners; but the levels picture is meant to articulate some metaphysically "tighter" connection—the lower levels in some sense *constitute* the higher levels. The relationship between statistical mechanics and thermodynamics should not be assimilated to that between the past and the future. The first and second ideas move in the direction of a tighter connection, but go too far: the relationship between the levels is discovered, not stipulated, and isn't a mere matter of meaning. What is wanted is a metaphysical not semantic relationship, but a tighter one than "material".

When I was in graduate school, a certain view was common of the available options for connecting distinct properties or subject matters, and more generally, for giving a philosophical account of something. The main two were definitions and synthetic necessities. One could say "x is good $=_{df} x$ causes pleasure", where the '$=_{df}$' was understood to be, in some sense, a matter of ordinary meaning. Or one could deny that 'good' can be defined (in the $=_{df}$ sense) but still hold it to be necessarily true that, for instance, anything that causes pleasure is good. A few other options were recognized (though not all of them were regarded as appropriate for philosophical accounts): extensional, nomic, and apriori. These connections can be ordered by the tightness of the connection:[16]

	extensional	$\forall x(Fx \rightarrow Gx)$
	nomic	$\boxed{\mathbb{N}}\forall x(Fx \rightarrow Gx)$
tighter connections	modal	$\Box\forall x(Fx \rightarrow Gx)$
	apriori	$\boxed{\mathbb{A}}\forall x(Fx \rightarrow Gx)$
	analytic	$Fx =_{df} Gx$

But it is natural to think that some further connection, intermediate in tightness between the modal and apriori, must be recognized. The connection between levels is metaphysical, not apriori; but it's tighter than a merely modal connection, for as we saw, domains can be modally connected even when there is no "constitutive connection", as was illustrated most dramatically by the Spinozistic view that all truths are necessary. Also, as Fine (and Jaegwon Kim (1990, section 4) before him) has emphasized, modal connections are not asymmetric. Its being necessary that all Fs are Gs leaves open that it's also necessary that all Gs are Fs; A supervening on B (in various senses) leaves open that B might also supervene on A. But, Fine argued, the levels picture (not his phrase) demands an asymmetric connection between lower and higher levels. Fine puts all this well, in a discussion of how to understand materialism about the mind:

It will not do, for example, to say that the physical is causally determinative of the mental, since that leaves open the possibility that the mental has a distinct reality over and above that of the physical. Nor will it do to require that there should be an analytic definition of the mental in terms of the physical, since that imposes far too great a burden on the [materialist][17]. Nor is it enough to require that the mental should modally supervene on the physical, since that still leaves open the possibility that the physical is itself ultimately to be understood in terms of the mental.

The history of analytic philosophy is littered with attempts to explain the special way in which one might attempt to "reduce" the reality of one thing to another. But I believe that it is only by embracing the concept of a ground as a metaphysical form of explanation in its own right that one can adequately explain how such a reduction should

[16] The ordering is oversimplified, in light of, for instance, the contingent apriori (Kripke, 1972). Also, to facilitate comparison with ground, the diagram lists "conditional" connections (except $=_{df}$); but one could instead consider biconditional connections: $\forall x(Fx \leftrightarrow Gx)$, $\Box\forall x(Fx \leftrightarrow Gx)$, etc. Correspondingly, one could consider a biconditional groundlike concept $\Leftrightarrow$, intermediate in tightness between modal and apriori equivalence. It would be biconditional in that $A \Leftrightarrow B$ would imply $\Box(A \leftrightarrow B)$, but nevertheless be asymmetric (as ground is normally held to be): if $A \Leftrightarrow B$ then $B \not\Leftrightarrow A$.

[17] Fine says "anti-realist", but he uses that word idiosyncratically; my substitution is arguably equivalent.

be understood. For we need a connection as strong as that of metaphysical necessity to exclude the possibility of a "gap" between the one thing and the other; and we need to impose a form of determination upon the modal connection if we are to have any general assurance that the reduction should go in one direction rather than another.

Fine (2012, pp. 41–2)

Thus the relation between facts at different levels is naturally taken to be ground: lower-level facts ground higher-level facts.[18]

1.7.2 *Wilson's challenge*

Jessica Wilson has argued that ground is in fact useless in philosophy. Consider its putative use in articulating "naturalistic" positions. According to Wilson, the bare claim that the mental, say, is grounded in the natural is neutral over a range of more specific positions involving more specific metaphysical relations such as type identity, token identity, functional realization, part–whole, and so forth. (Wilson calls the generic grounding relation Grounding with a capital G, and calls the specific metaphysical relations grounding-with-a-lowercase-g relations.) She says:[19]

Hence it is that naturalists almost never rest with the schematically expressed locutions of metaphysical dependence, but rather go on to stake out different positions concerning how, exactly, the normative or other goings-on metaphysically depend on the naturalistic ones.

On Wilson's view, then, grounding (i.e. Grounding) claims have no point; only the more specific claims are of interest.

Someone who viewed ground as a sort of super-added metaphysical force, so that facts about grounding are themselves fundamental facts not grounded in any further facts, would not agree that grounding claims are neutral over more specific positions; the grounding claim itself would count as another one of those specific positions. This isn't a very attractive view of ground though (we'll discuss it further shortly).

In my view Wilson is importantly right about something. When we attempt to say what is ultimately going on in some domain, metaphysically speaking, we don't stop with a claim of ground. We don't just say that the mind is grounded in the body and leave it at that (setting aside the super-added force conception of ground). Instead, as Wilson says, we go on to say something more specific about the connection between the mind and the body. I will argue later (sections 2.3 and 3.11) that, for this reason, neither ground nor essence is suitable as a tool for articulating the kinds of structuralist theses that are at issue in this book. (Thus the proper postmodal tool for our purposes is fundamentality.)

[18] I don't mean to suggest that there is any simple definition of the levels hierarchy in terms of ground, nor that all facts can be partitioned into levels.

[19] Wilson (2014, p. 546). Bennett (2017) and Kovacs (2017; 2018) make related claims, though Bennett holds that grounding claims—or rather, generic building claims, in her terms—have a role to play in metaphysics. See also Koslicki (2015).

But in other contexts, a less specific, more neutral claim is exactly what we want; thus I think that Wilson's critique of ground goes too far. We *do* stop with neutral claims of grounding, for instance, when we're stating overarching positions like physicalism or naturalism, or saying what makes more specific positions count as instances of physicalism or naturalism.

Why care about stating such overarching positions? Their usefulness is in their epistemic role. Such sweeping doctrines are epistemically important even if they're unspecific and hence in a sense metaphysically superficial. Take the case of consciousness. Physicalists work very hard to try to show that consciousness is somehow a physical phenomenon. They begin by exploring one sort of way to ground consciousness in the physical, but if that doesn't work, they try another way. Why do they stick to this path; why don't they just give up and concede that consciousness is a wholly nonphysical phenomenon? It's because they take themselves to have very good evidence that *everything* is grounded—in one way or another—in the physical. They think that the many cases in the history of science in which various phenomena that initially seemed not to be physicalistic were subsequently shown to be grounded in the physical collectively support a sweeping doctrine of physicalism, to the effect that *all* phenomena are grounded in the physical.

This line of thought essentially uses the general notion of ground, and cannot be reconstructed using any more specific relation, since different specific relations may be at issue in the different cases; chemistry, biology, and geology may be based in physics in different ways.[20] Thus the neutrality of ground over more specific metaphysical relations is essential to its epistemic role. (Modality also shares this neutrality.) When we're trying to get to the metaphysical bottom of things, we go deeper than ground. But in certain epistemic contexts it's important not to do this.[21]

To take one other example, anyone who accepts talk of fundamentality will want to state some sort of "completeness" principle, to the effect that all facts "rest" in some way on the fundamental; and it is natural to explicate this resting-on in terms of ground. One completeness principle, for instance, would say that any fact that involves any nonfundamental concept must be grounded in facts that involve only fundamental concepts.[22] Such a completeness principle is not a distinctive statement about what fundamental reality is like. It is rather a constraint on any

---

[20] Since property identity might be one of the ways (in the case of chemistry, for instance), the argument ought to employ Fine's (2012) notion of weak ground.

[21] Schaffer (2009) has stressed the value of ground in preserving a role for traditional metaphysical disputes given a Moorean respect for common sense (although see Sider (2013, sections 2, 4) against Mooreanism). This is another case in which the metaphysical neutrality of ground is essential: the Moorean demand is that one's fundamental metaphysics be capable of grounding *in one way or another* common sense, not that it ground common sense in one particular way. (See also Fine's (2012, p. 41) discussion of the importance of ground to the project of "critical" metaphysics.)

[22] Compare Sider (2011, section 7.5).

proposed inventory of the fundamental concepts: the inventory must be rich enough to accommodate all phenomena. To state such a constraint it is important *not* to be specific about the exact nature of resting-on, since different phenomena can rest upon the fundamental in different ways.

1.7.3 *Grounding ground*

Are facts about ground themselves grounded? Or are they ungrounded, as the "super-added force" view would have it?

For many grounding facts, such as those connecting levels, there is a powerful argument that they must be grounded. Levels-connecting grounding facts always involve higher-level concepts, because the higher-level fact getting grounded will involve such concepts; and surely no ungrounded fact involves a higher-level concept. Any fact of the form 'X grounds the fact that I am in pain', for instance, involves the property of being in pain, and hence is surely not ungrounded.[23]

The only exception to this sort of argument would be grounding facts involving only fundamental concepts, such as the fact that some particular e's being charged grounds the fact that something is charged. But even these facts seem unlikely candidates to be ungrounded; for why posit a new "super-added" force without good reason?

Thus grounding facts themselves have grounds. What grounds? I doubt there is any simple story to be told here, just as there is no simple story in general to be told about how higher-level facts are grounded. Ground, after all, is itself a high-level notion (assuming we reject the super-added force view), and one of the main reasons to accept ground in the first place is to allow for higher-level facts to depend on lower-level facts in complex ways that may not be accessible to us a priori. But we can make a good guess at the kinds of facts that help to ground grounding facts: patterns in what actually happens, modal facts, facts about the form or constituents of the grounding fact in question, metalinguistic facts, facts about fundamentality, and even (according to some friends of grounding though not me) certain grounding facts involving only fundamental concepts.[24]

1.8 Fundamentality

The final postmodal concept is that of fundamentality.[25] Actually there are two concepts worth distinguishing: that of a *fundamental fact*, and that of a *fundamental concept*.[26] The fundamental facts are those ground-level facts on which everything else rests. The fundamental concepts stand for the ultimate "building blocks" of the world, which "carve reality at its joints" and give it its fundamental structure.

Which, if either, of these two notions is more basic? On one view, the fundamental concepts may be defined as those standing for constituents of fundamental

---

[23] See Sider (2011, sections 7.2, 7.3, 8.2.1).

[24] See Sider (2020) for a fuller discussion of the issues in this section.

[25] The issues in this section are more fully discussed in Sider (2011).

[26] Some also speak of fundamental *individuals*; but see Sider (2011, 8.4–8.7).

facts. I myself prefer to leave 'fundamental concept' undefined (indeed, in my view, concept-fundamentality is itself a fundamental concept). In either case, the fundamental facts might be defined as those lacking grounds, or else taken as undefined. We can remain neutral on such issues.

Although concept-fundamentality is akin to Lewisian naturalness, it is more general in a certain way. There is little difference when it comes to concepts expressed by predicates: we may speak either of the fundamentality of the concept of being 1 g in mass, or of the naturalness of the property of being 1 g in mass. But we may (in my view) speak of fundamentality for concepts signified by expressions in other grammatical categories, such as sentence operators and quantifiers; and it is unclear that Lewisian naturalness can apply in such cases: naturalness for Lewis is a feature of properties and relations, and it is unclear whether the metaphysical function of operators and quantifiers is to stand for properties and relations. Just as Lewis would articulate the view that reality has fundamental "mass structure" by saying that mass properties (or relations) are natural, I would articulate the view that the world has fundamental ontological, or modal, or disjunctive structure by saying that the concepts expressed by quantifiers, modal operators, or the sentence operator 'or' are fundamental concepts. (This use of the term 'structure' has nothing to do with structuralism. The question is rather whether anything about mass, ontology, modality, or disjunction is woven into the ultimate fabric of reality, so to speak.) A concept—whether expressed by a predicate or no—is fundamental if and only if it plays a role in articulating the world's fundamental structure, if and only if it stands for one of reality's ultimate building blocks.

(Although the terminology may suggest otherwise, whether a concept is fundamental is not a matter of its place within our conceptual scheme, nor is it a matter of anything else about us; it is, rather, a purely worldly matter. The reason for speaking of fundamentality in terms of concepts is in large part to facilitate the generalization to logic, where the existence of entities standing to logical words as properties stand to predicates is contentious. See Sider (2011, chapter 6).)

There are certain abstract similarities and differences between the various postmodal notions. First, ground and fundamental facthood are *fact-level* (or propositional) whereas fundamental concepthood is *sub-factual* (or sub-propositional): it is entire facts that ground and are grounded, or are fundamental facts, whereas it is components of facts—or rather, their corresponding concepts—that are fundamental concepts. Essentialist claims $\Box_{x_1,x_2,...}A$ are partially fact-level (A) and partially subfactual ($x_1,x_2,...$). Second, ground is *comparative*, in that grounding claims involve multiple facts (one or more facts are said to ground another), whereas both fundamental concepthood and fundamental facthood are (on my usage anyway) *absolute*: fundamentality is fundamentality simpliciter—absolute fundamentality.[27] Essentialist claims $\Box_{x_1,x_2,...}A$ can be regarded as comparative: the

---

[27] I'm open to various concepts of relative fundamentality, but the more fundamental (!) concepts of fundamentality, in my view, are the absolute ones. See Sider (2011, section 7.11).

natures of $x_1, x_2, \ldots$ are said to give rise to A. However, the relevant facts about the natures of $x_1, x_2, \ldots$ aren't specified in the essential claim; indeed, there is no commitment to any such facts being specifiable. (We will return to this.)

Because it is comparative, there is a sense in which ground is a richer notion than either sort of fundamentality. Ground can be used to make assertions about high-level subject matters, and about how high-level matters relate to the lowest level, whereas (absolute) concept-fundamentality concerns only the lowest level. Moreover, as we saw, fact-fundamentality can apparently be defined in terms of ground, whereas there is no simple definition of ground in terms of fact-fundamentality. Thus the concepts of fundamentality cannot "go it alone" for certain philosophical endeavours. But for certain purposes this austerity of fundamentality can be welcome. Focusing exclusively on what is fundamental might be deemed appropriate if what one is giving an account of is itself a fundamental matter.

This austerity comes with a price. The absoluteness of fundamentality encodes a presupposition: that there is such a thing as an absolutely fundamental level. (It is of course not presupposed that we have knowledge of that absolutely fundamental level.) I defend this presupposition in Sider (2011, section 7.11). (Among other things, I point out that accepting the existence of absolutely fundamental concepts does not require the existence of mereological or spatiotemporal atoms; and I resist the idea that metametaphysical theorizing ought to be neutral about "first-order" metaphysical questions.) Still, the presupposition is a substantive one. Although some of what I will say about absolute fundamentality could be restated using a notion of relative fundamentality, much of it could not.

Let us finally discuss epistemology. How should we form beliefs about what concepts are fundamental?

Realist epistemology of science generally stresses the super-empirical virtues, notably simplicity of various sorts; and simplicity is in my view a central part of the epistemology of concept fundamentality.

One sort of simplicity, call it ideological parsimony, concerns the number and nature of undefined concepts: fewer and "simpler" concepts are better. Another sort concerns laws: a theory is better when it contains powerful yet simple laws, where the simplicity of a law corresponds to something about its syntax when stated using the theory's undefined concepts.[28] Frank Arntzenius's book *Space, Time, and Stuff* is a wonderful recent example of inquiry into the fundamental metaphysics of science that gives pride of place to simple and powerful laws. Arntzenius writes that:

... our knowledge of the structure of the world derives from one basic idea: the idea that the laws of the world are simple in terms of the fundamental objects and predicates. In particular, what we can know and do know about the way things could have been—what we can know and do know about the metaphysical, and physical, possibilities—derives

---

[28] Yet another sort is quantitative parsimony, positing fewer individuals; but in my view this is relatively unimportant. See section 3.14.1.

from our knowledge of what the fundamental objects and predicates are, and what the fundamental laws are in which they figure. I argue that it is bad epistemology to infer what the fundamental objects, predicates, and laws are on the basis of intuitions as to what is, and what is not, possible.

(2012, p. 1)

Notice how distinctively postmodal this is. Modal beliefs—about fundamental reality anyway—are epistemically downstream from nonmodal beliefs about the way reality is, and these nonmodal beliefs should in large part be determined by considerations involving laws (and also ideological parsimony, in my view). This epistemology will play a leading role in Chapters 3 and especially 4.

There are difficult questions about each sort of simplicity. Ideological parsimony is not just a matter of counting fundamental concepts, for instance, nor is simplicity of laws just a matter of measuring the length of their statements. But this is neither unexpected nor worrisome. Like all norms, epistemic norms are a high-level phenomenon, no doubt vague, perhaps somewhat contextual, and perhaps even incoherent in some cases.

Ideological parsimony is "negative", generating reasons against accepting concepts as fundamental. Simplicity of lawhood, on the other hand, is "positive", generating reasons for accepting concepts, when they are needed to formulate simple and powerful laws. There are laws-based negative maxims as well, such as not to posit fundamental concepts that aren't needed to state the laws, and not to posit fundamental concepts when simpler and equally powerful laws could be based on alternates, but perhaps these follow from ideological parsimony plus a purely positive simplicity-of-laws principle.

(A further positive epistemic force is not simplicity-based: the requirement that the fundamental concepts be "complete", that we posit enough fundamental concepts to capture all of the phenomena (section 1.7.2). This overlaps the simplicity of laws, insofar as the laws are among "the phenomena". Another epistemic force, which will be important in Chapters 4 and 5, isn't neatly classifiable as positive or negative: avoiding arbitrariness and artificiality. But there is surely more to the story.[29])

Realism about fundamental concepts and simplicity-based realist epistemology of science are made for each other. First, the realist about fundamental concepts is ideally placed to accept a simplicity-based realist epistemology. The ur-idea of this epistemology is that we are a priori entitled to expect the world to be simple. But it is a point familiar from Nelson Goodman (1955a, chapter 3) that simplicity

---

[29] For instance, John Hawthorne pointed out that it is unclear whether anything in the epistemology sketched so far counts against the idea that there is a single fundamental concept which completely specifies the nature of the whole universe as it actually happens to be. (I say "unclear" because of the requirement that the fundamental concepts be individually simple, but that is a pretty elusive requirement.) Against this idea I would invoke an additional epistemic principle, a preference for inventories of fundamental concepts that enable a complete account, not only of what actually happens, but also of the space of possibilities for what could happen. See Sider (2008a).

judgements depend on which concepts are deemed relevant to simplicity; a search for simple laws will lead in different directions, depending on which of 'all emeralds are green' and 'all emeralds are grue' is regarded as more simple. A realist about fundamental concepts recognizes an objective division amongst concepts, on which can be founded an objective simplicity-based epistemology.

Conversely, it's very natural for a realist about fundamental concepts to think that parsimony and simple-yet-powerful laws are epistemically important, provided she's a scientific realist anyway. For the realist about fundamental concepts believes in worldly distinctions corresponding to differences in these kinds of simplicity; and they seem like an exact match for the intuitive basis of realist thinking about theory choice, which is that the world is a priori likely to be simple, or that we are entitled to assume that it is.

It might be thought that only defenders of antireductionism about laws of nature—such as Armstrong (1983) and Maudlin (2007)—should centre their epistemology on simplicity of laws. Reductionists—like Lewis (1994)—don't think that laws are part of fundamental reality; and why should we expect simplicity in reality's derivative aspects? But this neglects a distinction between *laws* and law*hood*. To illustrate with Newton's dynamics: the law is just the fact that $F = ma$—that is, the fact that F for any object is in fact identical to $m \cdot a$ for that object—whereas the fact about lawhood is that the former fact indeed counts as a law. It is lawhood that reductionists think is metaphysically derivative: $F = ma$ counts as a law because of how this general fact fits into larger patterns (according to Lewis's particular form of reductionism anyway). But the law itself, the fact $F = ma$, is not derivative in this way; that fact concerns fundamental reality just as much for reductionists as for antireductionists. The laws-centric epistemology is as reasonable for a reductionist as for an antireductionist; an a priori bias towards simple patterns is as reasonable as an a priori bias towards simple robust-laws; each is a precisification of the vague bias towards the world being simple.[30]

---

[30] Another threat to the laws-centric epistemology might be thought to derive from Hicks and Schaffer's (2017) argument that fundamental laws need not be about fundamental properties. Newton's dynamical law, for example, is on their view about the nonfundamental property of acceleration. I am inclined to reply that the *metaphysically* fundamental law is not expressed by '$F = ma$', but rather by a more complex statement in which defined quantities like acceleration are replaced by their "definiens", so that only fundamental quantities appear (position, time, mass, and component force, perhaps). But Hicks and Schaffer object that this would be inappropriate meddling with physics; the textbook statement shouldn't be ruled out by a metaphysics of lawhood (sections 3.2, 4.3). I myself don't mind a *bit* of meddling: the metaphysician's conception of the law needn't match textbook statements so long as a reasonable methodology of the former can be given which isn't too detached from empirical methodology. But the issues are complex and can't be settled here. In any case, even if their argument is correct, there is no threat to the laws-centric epistemology, since that epistemology could be understood as merely requiring a bias towards simple statements about fundamental properties, whether or not true statements of that sort count as fundamental laws.

1.9 Apology

Some of the issues we've begun to discuss will strike some people as being overly "metaphysical", so let me close this chapter with an apology for the place of this sort of metaphysics in the philosophy of science. I myself think that a self-critical version of the "too metaphysical" reaction is important though probably wrong; but an uncritical version is indefensible.

Textbook statements of physical theories are often regarded as not being themselves foundationally adequate. Maybe they include some equations plus remarks about how to use the equations to make predictions, and nothing else. So we try to give a foundational account of the theory, to make clear "what the theory is telling us about the world".[31]

Whether a proposed foundational account is perceived as adequate is highly sensitive to the concepts in which the account is cast—the account's metaphysical tools. The question of what tools are appropriate is often left implicit, but it is itself substantive, central, and difficult. The uncritical reaction I deplore is simply presupposing one particular set of preferred tools without recognizing the substantive nature of the presupposition.

After all, consider someone just digging in and reiterating the textbook statement. 'You ask what Newton's dynamics is saying about the world? That's completely clear: it's saying that $F = ma$; what's the problem?' Here we want to object that this is not yet an adequate foundational statement. But the mere fact that the defender of the textbooks has just reiterated the theory in the original terms is not itself problematic. One can't keep recasting theories in other terms indefinitely.

Foundational accounts often include an explicit statement of the theory's *ontology*. For instance, in the case of classical mechanics we might make explicit the postulation of points of substantival Gallilean space-time and particles occupying that space-time. Here the challenge 'but what is *that* saying about the world' may be met either with bafflement or by simply reiterating the claim: 'well, it is saying that *there are* points of space-time and particles occupying them!'. At some point a statement of a theory is going to have to stand on its own.

Giving a theory's ontology is often considered a paradigm of an acceptable approach to foundations; indeed, many simply assume its acceptability. (Indeed, some seem to use the word 'ontology' to just *mean* 'what the theory says about the world'.) But that assumption is neither trivially correct nor universally shared. David Wallace, for instance, does not regard questions of ontology (such as whether it is three-dimensional space or some very high-dimensional space that fundamentally exists) as inevitably being good foundational ones[32]; and in metaphysics there are the ontological deflationists such as Eli Hirsch (2011) and Amie

---

[31] Clarifying a theory's metaphysics is just one possible goal of a foundational account. Other goals include clarifying its epistemology and what it says about measurement.

[32] See Wallace (2012, e.g. section 8.8). Wallace suggests structural realism as his preferred replacement for the status quo, but I offer him an alternative in section 5.6.1 of this book.

Thomasson (2007; 2015). According to some such philosophers, apparently incompatible claims about ontology might be nothing more than notationally different ways to get at the same reality. To them, the retort 'what's unclear? I've told you what there is!' is no better than the analogous digging-in of the defender of the textbooks.

According to the defender of the textbook, the concepts used in the textbooks were adequate metaphysical tools; according to the provider of the ontological foundational theory, the concepts of ontology ('there exist. . .') are adequate metaphysical tools. The question of which metaphysical tools really are adequate is deep, pervasive, difficult, and substantive. A theory stated using adequate metaphysical tools will hook up with what is objectively present in the world. It will not be in need of further metaphysical elucidation. Differences that are stated using adequate tools will be substantial differences, as opposed to "notational" or "merely conventional" differences. One won't have become "too metaphysical", by assuming structure that isn't really there. Thus the question of which tools are adequate is about how much structure reality has—a question that is as difficult and substantive as can be.

My own view is that a foundational theory must specify both a fundamental ontology and also fundamental concepts (including fundamental logical concepts, though this is more contentious and can mostly remain in the background). But the thing I want to stress here is the existence of the issue. Some carry on a dispute that presupposes a certain set of metaphysical tools, without acknowledging the presupposition. Others recoil from metaphysical issues that presuppose a certain set of metaphysical tools, without acknowledging that the recoil is a substantive reaction—why not those tools?—or that they themselves presuppose certain other tools as giving rise to genuine foundational questions. These are all substantive questions—foundational questions presupposing certain tools, and the questions of which tools are the right ones—and should be pursued simultaneously since there is two-way influence between the questions, as I hope to illustrate.

2

Nomic Essentialism

Our overarching theme is how metaphysical inquiry is sensitive to one's chosen metaphysical tools, especially in the metaphysics of science, and extra especially in the evaluation of structuralist claims. In this chapter we turn to the first of three case studies to develop this theme: the relation between scientific properties (and relations) and the laws of nature.[1]

Are scientific properties "independent" of the laws of nature in which they figure? "Quidditists" say yes. "Nomic essentialists" say no: properties are bound up with laws.

Actually, the literature has mostly discussed the relation of properties to causation, dispositions, and powers, rather than to laws.[2] But I'm going to talk about laws since I want to explore the issue as it arises at the fundamental level, and laws, I think, are more likely to be fundamental than causation or dispositions or powers.

Arguments in favour of nomic essentialism often involve "exchanging nomic roles". A property's nomic role specifies which laws of nature it obeys. If properties are "independent" of the laws, as quidditists think, then it should be possible for a property to have a different nomic role—to obey different laws from those it actually obeys. It should even be possible for a pair of properties to exchange their nomic roles, so that, for example, charge obeys the kinds of laws that mass actually obeys, while mass obeys the kinds of laws that charge actually obeys. Nomic essentialists then object to this consequence in various ways. Some object that if role-exchanging were possible, then we couldn't know certain things that we obviously do know, such as that I am more massive than a mouse. For a world in which charge and mass have exchanged roles would appear exactly the same to us, but I might not be more massive than a mouse in that world. Others object that a world in which charge and mass have exchanged roles is a "distinction without a difference".

For the record, I don't myself find either argument convincing. The possibility of role-exchanging does not threaten our possession of knowledge in various ordinary senses of 'knowledge', and there is no reason to suppose that we possess

---

[1] Those less interested in purely metaphysical issues may wish to skip ahead to the next chapter, although section 2.3 introduces an important recurring theme.

[2] E.g. Shoemaker (1980); Swoyer (1982); Cartwright (1999); Bird (2007a). But see Hawthorne (2001).

knowledge in any of the extraordinary senses in which such knowledge would be precluded.[3] As for the metaphysical argument, it seems either to rely on an observational criterion for being a distinction without a difference (the possibilities "look the same"), in which case it has no appeal to anyone who takes seriously the kinds of metaphysical issues discussed in this book (or indeed, anyone who believes in unobservables in science), or else is nothing more than the question-begging assertion that there is in fact no fundamental difference between role-exchanging worlds.[4]

But our focus here will be on how nomic essentialism should be understood, not on whether it is true. The question of whether properties are "independent" of the laws of nature needs to be made precise, and how this is done depends on one's metaphysical tools. The issue has mostly been approached from a modal perspective, with the central questions being whether a property could have obeyed different laws, or whether there are possible worlds in which properties exchange nomic roles. But it's natural to be dissatisfied with a merely modal formulation. If properties can't have different nomic roles, this ought to follow from some deeper postmodal claim about the actual metaphysical relationship between properties and laws. Such a claim will likely have modal implications, so addressing the modal issues remains important, but the deeper claim is the one of ultimate interest, or so I will assume.

In fact, nomic essentialism is hard to formulate postmodally; that is the main thesis of this chapter. Quidditism, on the other hand, is perfectly straightforward from a postmodal point of view. A committed nomic essentialist might say: so much the worse for postmodal metaphysics. My own reaction is rather: so much the worse for nomic essentialism. Either way, our overarching theme will have been reprised; the tools of metaphysics matter. We will also begin to encounter the obstacles confronting "structuralist" views in general.

It is natural to expect the postmodal formulation of nomic essentialism to somehow downgrade scientific properties, relative to the laws of nature. Nomic structure is primary; the nodes in that structure are secondary. Now, the most obvious way to downgrade the metaphysical status of entities is to eliminate them, to deny that they exist. But in the present case that would seem to be incoherent. Nomic structure consists of laws of nature, which are facts about scientific properties; how then could nomic structure be upheld if the existence of

---

[3] See Langton (2004), Schaffer (2005), and also Cross (2012a, pp. 133–6) and Hawthorne (2001, section 3). Dasgupta (2015b, p. 610) has pointed out in a different context that alternate possibilities for role-occupation differ in important ways from familiar sceptical hypotheses, so it may be that Langton and Schaffer over-emphasize the analogy to the traditional problem of scepticism. (See also Dasgupta (2015a, p. 474) for an account of the kind of ignorance implied by the possibility of role-exchanging.) But I don't see this as undermining the core of the reply.

[4] Another argument I don't find convincing is that 'science finds only dispositional properties' (Blackburn, 1990, p. 63). What science gives us is an inventory of scientific properties and a specification of their nomic roles, not the metaphysical relationship between the two.

properties is denied?[5] Some structural realists, whom we will discuss in Chapter 3, do make just such a paradoxical denial in the case of individuals, but in this chapter let us consider more conservative ways to downgrade properties relative to laws.

2.1 Nomic essentialism and ground

First let's look for a ground-theoretic formulation of nomic essentialism, and in particular for one saying that certain facts about scientific properties are grounded in laws. Such a formulation would be natural since it would privilege the pattern of law-like relations amongst properties over the individual properties themselves.

It's further natural to assume that such a thesis would take a more specific form, namely that whenever a property is instantiated, this fact is grounded in facts about laws. Thus whenever an object a has a property P, the nomic essentialist would accept a claim of the following form:

$$[\text{something about the laws}] \Rightarrow a \text{ has } P.$$

But what about the laws, exactly, will go on the left-hand-side? It must connect to the object a. But laws do not speak of particular objects by name, so to speak; they speak of objects via their properties. Moreover, the relevant property for a will surely be P. Won't this involve a circularity of ground?[6]

To be concrete, suppose the view has this form:

$$\exists p(\mathscr{L}(p) \wedge a \text{ has } p) \Rightarrow a \text{ has } P.$$

"a has P because a has some property with nomic role $\mathscr{L}$."

The formula $\mathscr{L}(p)$ expresses P's nomic role. It may be defined as the result of beginning with a sentence $\mathscr{L}(P)$ stating laws of nature in which P figures, and

---

[5] Although see sections 2.6 and 3.16.

[6] The argument is only a cousin of Russell's concern about things being "each other's washing" (1927, p. 325), on which see Robinson (1982, section 7.4); Blackburn (1990); Holton (1999); Hawthorne (2001); Bird (2007b); Lowe (2010); Cross (2012b). (Though see Whittle (2008), whose concerns are closer to mine.) The washing argument is about how properties are defined or "individuated"; mine is about how property instantiations are grounded. One quick comment about Bird's treatment of Russell's argument. At times he seems to construe dispositional essentialism in terms of essence. But if essence is understood in Fine's sense, so that the view is like the one to be considered in section 2.3, then it is unclear what the regress concern is. (If cycles of essence-involvement—"reciprocal essences" in Fine's (1994b) terms—are deemed problematic (Bird, 2007b, p. 516), the view could be that the laws hold in virtue of the essences of all properties collectively.) Really, though, Bird seems to construe dispositional essentialism modally, since his main response to the argument is to formulate dispositional essentialism as the claim that there are modally necessary and sufficient conditions, stated in terms of dispositional relationships, for being identical to a given property. But if the view is to be formulated as a merely modal thesis, then why such a restrictive one, which for example rules out possible worlds with symmetrical dispositional structures? Why not the weaker thesis mentioned by Hawthorne (2001, part 3), namely that worlds with the same dispositional patterns contain the same properties? It's not as if the stronger modal thesis yields a dispositional account of *what properties are* in any postmodal sense.

then replacing all occurrences of the name 'P' with the variable 'p'. (There are several choice-points here, on which we may remain neutral. Must $\mathscr{L}(P)$ include all of P's laws? Must it include a "that's-all" clause, saying that no other laws govern P? May properties other than P be mentioned by name in $\mathscr{L}(P)$, or must they be "ramsified out", so that $\mathscr{L}(P)$ says that there exist properties $q, r, \ldots$ that are connected by law to P in a certain way?)

Next make some standard assumptions about the logic of ground.[7] To make these assumptions we need a distinction between "partial" and "full" ground (Fine, 2012, section 1.5). Full ground ($\Rightarrow$) is the notion of ground we've been discussing so far. A partial ground is a "part" of a full ground: one fact partially grounds ($\rightsquigarrow$) another fact when the first, perhaps together with other facts, fully grounds the second.[8] The assumptions are then these:

For all y, if $A(y)$ then: $A(y) \Rightarrow \exists x A(x)$.

("Existentials are fully grounded in each of their true instances.")

If $A \wedge B$ then: $A \rightsquigarrow (A \wedge B)$, and $B \rightsquigarrow (A \wedge B)$.

("Conjunctions are partly grounded in their conjuncts.")

If $A \Rightarrow B$ then $A \rightsquigarrow B$.

("Full ground implies partial ground.")

$\rightsquigarrow$ is transitive and irreflexive.

The problem is then as follows. We've supposed that a's having P is fully grounded in the fact that $\exists p(\mathscr{L}(p) \wedge a \text{ has } p)$. A true instance of this existential is $\mathscr{L}(P) \wedge a \text{ has } P$; and so the existential is grounded fully and hence partially in the instance (existentials...); but the instance is partially grounded in the fact that a has P (conjuncts...), which given transitivity violates irreflexivity.

2.2 Other grounding claims: existence, identity

Given the previous section, nomic essentialists might back away from the claim that facts about the *instantiation* of properties are grounded in the laws, and say instead that some other facts about properties are grounded in the laws.

[7] The assumptions are implicated in certain ground-theoretic paradoxes (Fine, 2010a; Correia, 2013); I presume that any needed revisions to them would not undermine their use here.

[8] This definition of partial ground implies a "remainder" principle that if A partially grounds B then there are facts which include A and together fully ground B. There might be interest in denying the remainder principle (e.g. to accommodate Williamson's (2000, chapter 3) "prime" concepts or Rosen's (2017) puzzle about normative grounding for non-naturalists) which would require rejecting the definition. But my argument doesn't require the remainder principle, only its converse.

(In my view they should *not* back down in this way. After all, isn't it part of their core idea that we simply can't make sense of what it is for a given thing to have a certain mass, for example, independently of the role that mass plays in the laws of nature? Again: their core idea, I would have thought, is that the properties themselves are to be accounted for in terms of the laws; but surely the way one "accounts for" a property is to give an account of its instantiation. But in any case, let's look at where backing down leads.)

A nomic essentialist might, for instance, say that it's the *existence* of a property, and not facts about its instantiation, that is grounded in the laws. Perhaps facts about the property's instantiation are fundamental, on this view, having nothing to do with the laws. The laws enable the property to get its foot in the door of being; but once it's in, the property no longer needs the laws in order to be instantiated by particulars (though of course these instantiations are guided by the laws, in a sense dependent on the particular view of lawhood adopted by the nomic essentialist).

But what kind of fact about the laws will ground a property P's existence? It surely can't be the fact $\mathscr{L}(P)$, the fact that P itself plays a certain role in the laws of nature, for that fact "presupposes" P's existence. How could P's possession of *any* feature ground its own existence? Nor can it be the fact that $\exists p \mathscr{L}(p)$, the fact that some property plays that role, since this fact is grounded in its instance $\mathscr{L}(P)$ which presupposes P's existence. The intuitive problem here is this: the idea was to ground the existence of P in "the laws", but the relevant facts about the laws involve P, or are grounded in facts that involve P, and hence cannot ground P's existence.

Instead of saying that the laws ground facts about property instantiations or existence, the nomic essentialist might instead say that they ground facts about property identitites. Nomic essentialists in fact do say things like this: 'the *identity* of a property involves the laws'.[9] So perhaps the nomic essentialist could claim:

$$\mathscr{L}(P) \Rightarrow P = P.$$

But for one thing, I doubt identity facts have grounds.[10] For another, even if they do, surely the ground for an object's identity with itself should have a very general character, so as to be applicable to objects of any sort. The idea that an object's self-identity is grounded in that object's existence would be general in this sense; the proposed claim that $P = P$ is grounded in $\mathscr{L}(P)$ would not be since it is applicable only in the case of the self-identity of a scientific property. And anyway, the nomic essentialist idea that properties are tied to their nomic roles seems distant from the idea that facts about the laws ground such facts as $P = P$. To verify this impression, consider the modal implication of the claim that $\mathscr{L}(P) \Rightarrow P = P$: it is $\Box(\mathscr{L}(P) \rightarrow$

[9] See, for example, Bird (2016, p. 4); Mumford (2004, p. 151).

[10] Not identity facts for entities that fundamentally exist anyway. But see Burgess (2012); Fine (2016); Shumener (2020).

$P = P$). (I assume the usual view that ground implies necessitation.) This modal claim does nothing to block the possibility of properties exchanging nomic roles. Where $\mathscr{L}$ and $\mathscr{L}'$ are the nomic roles of properties P and P', respectively, so that $\mathscr{L}(P)$ and $\mathscr{L}'(P')$, for all the modal claim says there might be a possible world in which $\mathscr{L}(P')$ and $\mathscr{L}'(P)$—the modal claim's meager implication about this world is just that $P = P$. The original ground-theoretic proposal, in contrast, does block role-exchanging. That proposal was that $\exists p(\mathscr{L}(p) \wedge a \text{ has } p) \Rightarrow a \text{ has } P$, whose modal implication is that $\Box(\exists p(\mathscr{L}(p) \wedge a \text{ has } p) \rightarrow a \text{ has } P)$. Thus in a putative world in which P exchanges nomic roles with P', a and other objects that instantiate the property that has P's actual role—namely, $\mathscr{L}$—would still be instantiating P.

It wouldn't help to change the grounding claim to:

$$\exists p \mathscr{L}(p) \Rightarrow P = P.$$

The corresponding modal claim is then $\Box(\exists p \mathscr{L}(p) \rightarrow P = P)$, which still allows the role-exchanging world in which $\mathscr{L}(P')$ and $\mathscr{L}'(P)$; its meager implication about this world is again just that $P = P$ (via the implication of $\exists p \mathscr{L}(p)$ by $\mathscr{L}(P')$).

A more promising suggestion would be to interpret talk of "P's identity" being grounded as involving an irreducibly generic or general concept of grounding, of the sort recently advocated by a number of writers.[11] The idea can be implemented by attaching the grounding operator to sentences with free variables; in these terms the most recent proposal could be amended to:

$$\mathscr{L}(p) \Rightarrow p = P.$$

(with p a variable). Thus instead of specifying the ground of the *fact* that P is identical to itself, we specify the ground of the *kind* of fact (or property of facts) of being a fact of identity with P. The usual view that ground implies necessitation has its analogue for general grounding: if $A(x) \Rightarrow B(x)$ then $\Box \forall x(A(x) \rightarrow B(x))$; thus the general-grounding claim implies that any property satisfying $\mathscr{L}$ must be P itself, which does preclude the possibility of exchanging nomic roles.

I myself am inclined to make certain complaints about this proposal, but they may have limited dialectical force. First, the involvement of the identity relation in the proposed claim of general grounding seems extraneous to the proper concerns of nomic essentialists. Those concerns involve charge, mass, and so forth, and not the identity relation at all. Second, the same reason for doubting that identity facts have grounds is also a reason to doubt that identity-with-a-given-thing has a ground—identity just seems basic. Third, it seems intuitively odd that laws play the limited grounding role of securing identities of properties but no role at all in the instantiations of those properties. I do have a further complaint which is considerably more forceful (I think), but discussion of it must await the next section.

---

[11] Fine (2016); Wilsch (2015); Glazier (2016); see also Schaffer (2017b, section 4.1) on functional metaphysical laws.

2.3 Nomic essentialism and essence

Given some of the roadblocks of the previous section, it is natural to ask whether the key nomic essentialist claim might be better understood as involving Fine's notion of essence. Fine often glosses the claim that an object is essentially thus-and-so by saying that being thus-and-so is part of that object's "identity". So perhaps the claim that a property's identity involves the laws should be understood not as a grounding claim involving the identity relation, but rather as a Finean essentialist claim:

$$\Box_{P,Q,\dots}\mathscr{L}(P,Q,\dots).$$

That is: 'the properties $P, Q, \dots$ are essentially such that $\mathscr{L}(P, Q, \dots)$', where $\mathscr{L}(P, Q, \dots)$ is the nomic role played by $P, Q, \dots$.

Notice an intuitive difference between the essence-theoretic formulation of nomic essentialism and formulations we'd been considering previously. The previous formulations all embrace a laws-to-properties "direction of influence", in one way or another. The initial formulation considered in section 2.1, for example, claimed that particular property instantiations hold in virtue of facts about laws. But, intuitively, the essence-theoretic formulation embraces the reverse, saying that laws spring from the natures of properties. For the Finean gloss on $\Box_{P,Q,\dots}\mathscr{L}(P,Q,\dots)$ is that $\mathscr{L}(P,Q,\dots)$ holds in virtue of the natures of $P, Q, \dots$.

This interpretation of nomic essentialism does not bring in the extraneous subject matter of the identity relation—a plus, I think. Also, the modal claim implied by this reading of nomic essentialism, namely that properties have their nomic roles necessarily (following Fine in assuming that essence implies necessity), is certainly close to the concerns of nomic essentialists. It rules out the role-exchanging world considered earlier: since $\mathscr{L}(P)$ holds in actuality, it must hold in every possible world; but in the putative role-exchange-world in which $\mathscr{L}'(P)$ and $\mathscr{L}(P')$ hold, $\mathscr{L}(P)$ does not hold if the roles $\mathscr{L}$ and $\mathscr{L}'$ are incompatible.[12] It also agrees at an intuitive level with many of the things that nomic essentialists say. As Alexander Bird (2007a, p. 2) puts it, '. . .laws are not thrust upon properties, irrespective, as it were, of what those properties are. Rather the laws spring from within the properties themselves.'

---

[12] Prima facie, the essentialist claim does not rule out everything in the vicinity of role-exchanging, such as the possibility of a given property being "replaced" by a distinct property with the same nomic role. There are subtle issues here, though. There is this argument, for instance: if there could have been some other property Q playing P's actual nomic role, then it could have been necessary that Q play this role, in which case Q must actually exist and play this role. There are various well-known ways to try to resist this argument, to be sure. In any case, an essentialist might supplement her view with the claim of general-grounding considered at the end of the previous section, which would rule out such possibilities. Thanks to Jonathan Schaffer here.

I suspect the essence-theoretic proposal will be the most popular I'll consider. But in my view it is very unsatisfying, even if true. It says that something "flows from the essences of $P, Q, \ldots$" without saying *how* that something flows.[13]

Our search for a formulation of nomic essentialism began with the postmodal conviction that a merely modal formulation is inadequate. Modal claims, such as that it would be impossible for a property to obey other laws, are not adequate statements of nomic essentialism on their own, according to this conviction. We need some deeper, more revealing metaphysical account of properties and laws from which the modal claims would follow.

But now, in the essentialist formulation, we are given the bald statement that properties' essences give rise to their laws, without any more substantive statement of what those essences are, without any account of what it is about properties or laws in virtue of which they are so tightly connected. This does not, it seems to me, scratch the postmodal itch any better than merely modal claims do.

To bring this out, begin with an ontology of fundamental properties which are metaphysically "rock-bottom", bearing no constitutive relations to anything else. Suppose further that we accept some robust, antireductionist account of laws—Armstrong's, perhaps—according to which the laws are also constitutively unrelated to other elements of the metaphysics. Laws and properties (and individuals, let's suppose) are merely "externally" related to one another. This is pure quidditism.

Now, suppose that without disrupting any of the elements in the picture so far, we simply *add* something: a modal fact that it would be impossible for any property to obey different laws. The addition is to be "external" to the elements introduced earlier: neither the internal nature of properties nor their laws are to be altered. From a postmodal point of view, this clearly would not be a legitimate form of nomic essentialism. Slapping some □s on the original quidditist conception of reality wouldn't remove the quidditism. It might be objected that modal facts just can't be added without altering the original facts intrinsically, since modal facts derive from the internal nature of the fundamental facts. Fine; but then the postmodal demand will be for an account of that internal nature. The modal claims would not themselves be nomic essentialism; rather, nomic essentialism would be a claim about the internal nature of properties and laws, from which the modal claims would follow.

Anyone on the postmodal train will, I take it, regard the previous paragraph as unobjectionable. But now consider the essentialist parallel. Suppose we simply add to the initial quidditist picture an essentialist fact, that it is essential to the fundamental properties that they obey their laws (that they stand in Armstrongian relations of necessitation, suppose). The addition again is to be external; neither

---

[13] Barker (2013) makes a similar complaint. At the end of his paper Barker suggests a general critique of holistic metaphysics that is akin to that of this book; and there are further points of contact between that paper and various parts of Chapter 3 of the present work.

the properties nor the laws are to be intrinsically altered. This addition wouldn't result in a satisfying form of nomic essentialism either. Slapping on $\Box_{P,Q,...}$s doesn't remove the quidditism any more than would slapping on unsubscripted $\Box$s. As before, it might be objected that the addition just couldn't be external since essentialist claims derive from the internal nature of the fundamental facts; but as before, this plays into my hands. For it would then seem that the proper statement of nomic essentialism should not be the essentialist claim, but should rather be some account of the internal nature of properties and laws that would give rise to the essentialist claim.

The gloss on a Finean essentialist claim $\Box_{x_1,x_2,...}A$ is that 'A holds in virtue of the natures of $x_1, x_2, \ldots$'. This can suggest a more informative account, that there are these things, natures, which are had by $x_1, x_2, \ldots$ and which explain why A holds. But it isn't as if one has been given an account of natures, or of how they give rise to the truth of statements. Ultimately, natures are given no more explicit articulation than: 'are such as to give rise to certain essential truths'.[14]

It might be conceded that the truth of the essentialist formulation of nomic essentialism would indeed be due to some distinctive facts about the natures of properties and laws, but claimed that this does not undermine its status as the correct formulation. That formulation, it could be claimed, is neutral over various more specific visions that all count as nomic essentialist.

For certain purposes this sort of attitude would be appropriate. Some important metaphysical questions are perhaps answered by saying which facts flow from things' natures, and by identifying the things in question, but without saying anything more about those things' natures. But the attitude would be out of place here, for two reasons. First, nomic essentialism is surely meant to be some more specific vision about the ultimate natures of properties and laws. And second, unless some more specific vision is articulated, it will remain unclear whether there is any sufficiently attractive specific vision of the natures of properties and laws that would underwrite the essentialist claim. After all, it is exactly that sort of specific vision that we have been struggling to find.[15]

---

[14] Compare Fine (1995, p. 273):

> Although the form of words 'it is true in virtue of the identity of x' might appear to suggest an analysis of the operator into the notions of the identity of an object and of a proposition being true in virtue of the identity of an object, I do not wish to suggest such an analysis. The notation should be taken to indicate an unanalyzed relation between an object and a proposition. Thus we should understand the identity or being of the object in terms of the propositions rendered true by its identity rather than the other way round.

[15] What if the nomic essentialist said that the difference between the quidditist and nomic essentialist conceptions of fundamental reality is nothing more than a difference in the identities of fundamental properties? Fundamental properties $P_1, P_2, \ldots$ are quidditist, fundamental properties $P_1', P_2', \ldots$ are nomic essentialist, and nothing can be said about this difference beyond that only the latter are such as to give rise to necessary or essential truths connecting them to laws. This strikes me as very unsatisfying in something like the way in which "quotienting" is unsatisfying; see section 5.5.

My point here is in the spirit of Wilson's (2014) critique of ground, that grounding claims are useless because they're neutral over more specific claims. I argued in section 1.7.2 that this neutrality of ground doesn't undermine its use for certain purposes, but the statement of nomic essentialism is one of those cases where I think that a point like Wilson's—applied to essence rather than ground—is right.

My objection has been that the essentialist formulation does no better than the modal formulation in articulating a metaphysically specific account of the natures of properties and laws. But it might be objected that some essentialist claims *can* play a role in articulating a specific account of their subject matter, namely those essentialist claims that are underwritten by real definitions.[16] Suppose, for instance, that it is essential to causation that an event c causes an event e if and only if e would not have occurred had c not occurred, and that this is underwritten by a real definition of causation as counterfactual dependence. This real definition is a specific and substantive account of causation. Indeed, it is a reduction of causation: the totality of facts about causation, given the real definition, is nothing more than a collection of facts about counterfactual dependence.

But not all real definitions—or anyway, not all claims that are alleged to be real definitions—can play this kind of role. The claim that {Socrates} essentially contains Socrates is sometimes claimed to be underwritten by a real definition of {Socrates} as the set which contains Socrates and only Socrates. But such putative real definitions of sets do not yield a substantive account of the nature of set-theoretic facts, in the way that the real definition of causation as counterfactual dependence yielded a substantive account of the nature of causal facts. In the latter case, "substitution" of real definiens for real definiendum in any fact about causation yields a fact that does not mention causation: substitution in the fact that c causes e yields the fact that if c hadn't occurred, e wouldn't have occurred. But such substitution in the fact that Socrates is a member of {Socrates} yields the fact that Socrates is a member of the unique set containing Socrates and only Socrates—a fact that mentions sets, and indeed, contains a quantifier ranging over a domain that contains the very object {Socrates}. (The definition of {Socrates} is in a sense "impredicative".) No substantive account of set-theoretic facts in general, or of particular sets like {Socrates}, is yielded by real definitions like that of {Socrates}, since substitutions using those definitions yield facts that concern the very set-theoretic structure and objects that were to be accounted for. Such real definitions certainly do not yield reductions. Thus the putative real definition of {Socrates} is quite different from the real definition of causation as counterfactual dependence (the latter, notice, is predicative).

Could a real definition underlie the claim that it is essential to properties $P, Q, \ldots$ that $\mathscr{L}(P, Q, \ldots)$, and if so, would it be like the real definition of causation or the real definition of {Socrates}? The only real definition fitting the

---

[16] Thanks to Shamik Dasgupta for discussion of this issue.

NOMIC ESSENTIALISM AND ESSENCE 33

bill would seem to be a real definition of $P, Q, \ldots$ as the properties $p, q, \ldots$ such that $\mathscr{L}(p, q, \ldots)$. But this is exactly parallel to the real definition of $\{\text{Socrates}\}$ as the set containing Socrates (it is impredicative in the same way): substitutions in statements about properties and their nomic roles via such definitions result in statements about the very same subject matter. If we want some illumination of properties and their connection to their nomic roles, it does not help to be told that those properties are defined as the properties that play those roles.

I have been demanding a more informative account of the nature of scientific properties that underwrites essentialist claims about them. But it might be objected that this demand is inappropriate since some clearly correct essentialist claims cannot be thus underwritten. It is essential to the concept of disjunction, perhaps, that it obeys the rule of disjunction introduction, but this is presumably not underwritten by any reductive or predicative definition of, or any other more fundamental account of, disjunction. (The impredicative real definition of disjunction as the operation satisfying certain inferential rules would not yield a more informative account, for the same reason that the definition of $\{\text{Socrates}\}$ did not.) All we can say is that disjunction is essentially such as to obey that rule.[17]

Essentialist claims sometimes *can* be given a more informative account, as in the example of the counterfactual definition of causation. One further example (this time one where the underwriting claim is more distant from the underwritten one): the fact that the ancestor relation is essentially transitive can be explained by a (predicative) real definition of being an ancestor: for x to be an ancestor of y is for x and y to be the first and last, respectively, in a finite series of objects in which each but the first is a parent of the previous. This definition implies that being-an-ancestor is transitive, since if x and y are connected by such a series and y and z are as well, then x and z are connected by a series consisting of the first two series concatenated. The objection must therefore be that even though some essentialist claims can be further explained in this way, there are other essentialist claims, such as the claim that disjunction obeys its customary inference rules, that need no further explanation.

But if the essentiality of nomic roles to properties is no more explicable than the essentiality of disjunction-introduction to disjunction, the essentialist formulation of nomic essentialism is barely an improvement on the modal formulation. The modal formulation says that it's necessary that if a property exists then it obeys its laws. The essentialist formulation is nearly identical, adding only that this necessity has its source in the properties, while conceding that nothing more can be said about why this is so. I suppose that pointing to the properties as the source of the necessity does head off certain concerns about the modal formulation, such as that the necessity could be due to a general feature of modality (the "Spinozistic view", for instance) having nothing to do with the properties. But the account still remains distant from the postmodal ideal, of a satisfying account of the structure

---

[17] Thanks to Alexander Bird for a helpful discussion here.

of actuality giving rise to the modal claim. We were seeking an improvement on the merely modal formulation; a "brute" essentialist formulation doesn't deliver it.

(Indeed, if a certain sort of reductionism about essence is true, the essentialist formulation wouldn't differ from quidditism! Suppose a thing's essence just consists in certain sorts of important facts about it. Which facts? Well, specifying them will be the task of the reductionist, but on a rather deflationary approach, the specification might be somewhat conventional. And in the case of a property, one of the important facts might be held to be the property's nomic role. Nomic essentialism would then amount merely to the claim that a property's nomic role is one of the specified important facts about that property—which is something that a quidditist could accept.

At the opposite end of the spectrum from conventionalist reductionism about essence is the view that essence is metaphysically basic. This would be like the "further force" view of ground mentioned in Chapter 1. Given that view, the essentialist articulation of nomic essentialism would at least amount to a statement about fundamental reality. Now, I don't myself like this view for essence any more than for ground. But even given the view's truth, the articulation of nomic essentialism that it enables remains unsatisfying. We may again compare it to a merely modal articulation, given the view that modality is metaphysically basic. In each case, a demand for an account of the underlying nature of properties and laws still seems appropriate. A metaphysically basic modal or essentialist connection between laws and properties would be "external" to laws and properties, whereas one would have hoped for an "internal" account, an account of the "innards" of one or the other illuminating the distinctive connection between them.)

I have been complaining that '$\mathscr{L}(P,Q,\dots)$ holds in virtue of the essences of $P,Q,\dots$' is insufficiently metaphysically revealing to count as the statement of nomic essentialism. I'd make the same complaint about ground-theoretic variants of that formulation, such as '$\mathscr{L}(P,Q,\dots)$ is grounded in the fact that $P,Q,\dots$ exist' (or in $P,Q,\dots$ themselves, if, as Schaffer (2009) thinks, particular entities can ground). *How* does the existence of some properties ground facts about their nomic role? What are the relevant features of the properties by virtue of which the putative grounding claim holds? (This complaint applies only to some ground-theoretic articulations of nomic essentialism. It does not apply, for instance, to the proposal considered in section 2.1, since that implies a distinctive claim about the most fundamental facts involving scientific properties.)

And, to complete an argument from the previous section, I would make the same complaint about the claim of generic grounding that playing role $\mathscr{L}$ grounds being identical to P, i.e. $\mathscr{L}(p) \Rightarrow p = P$. Like the formulations considered in the present section, this doesn't on its own yield any distinctive general account of fundamental reality, nor does any such theory with which it might be supplemented come to mind.

Claims of generic ground don't on their own constitute a distinctive general account of fundamental reality. What they can do is allow us to "place" nonfundamental matters within a given account of (more) fundamental reality. Suppose it given, as a fundamental fact, that there exists a certain mereological sum of subatomic particles with feature T_1; and suppose the generic grounding claim that T_1 grounds tablehood: $T_1x \Rightarrow Tx$. We can then say that the sum of particles is a table, and indeed that it's a table because it's a T_1. We have "placed" the feature tablehood on the pre-existing fundamental grid of particles and their mereological sums, by means of the generic grounding claim. But in order to do so, we needed that grid in the first place.

Can the generic grounding claim $\mathscr{L}(p) \Rightarrow p = P$ be used in this way to place particular scientific properties P within a pre-existing fundamental grid that, somehow, does justice to the nomic essentialist vision? We're back where we started: what is that fundamental grid? What is the nomic essentialist vision of fundamental reality? Also, this particular generic grounding claim concerns an *identity property*, the property of being-identical-to-P. So the kinds of grids in which it can "place" this property must contain a property possessing the feature $\mathscr{L}(p)$. This property will be the very property P! So it isn't as if the generic essentialist claim will enable some fundamental account to dispense with facts about particular scientific properties. At best it would seem to be a kind of add-on to what would otherwise appear to be the quidditist's grid: fundamental facts about the possession of properties by particular entities—a has P, b bears R to c, etc.—plus facts about laws, such as the fact that $\mathscr{L}(P)$. It's hard to view the addition of the generic grounding claim to this grid as an advance on the merely modal formulation of nomic essentialism. The initial grid contains nothing distinctive of nomic essentialism, and no hint of why exchanging of nomic roles should be impossible. The impossibility of role-exchanging is guaranteed only when the generic grounding claim is added.

The complaint I've been making throughout this section is akin to the standard postmodal complaint about modal formulations of metaphysical theses. The postmodal complaint that modal theses are insufficiently revealing of "metaphysical structure" is based on the hankering for a more explanatorily satisfying account than mere modal formulations offer. What we have seen is that even certain postmodal formulations leave us with the same hankering.

2.4 Nomic essentialism and fundamentality

Might nomic essentialism be formulated in terms of fundamentality, rather than ground or essence?

The debate between nomic essentialists and quidditists is, intuitively, over whether properties are independent of laws. This might suggest characterizing quidditism as the view that scientific properties are fundamental properties (or, in my official terms, that their concepts are fundamental concepts), and/or that singular facts involving their instantiation are fundamental facts. This would secure the requisite "independence" of properties, and nomic essentialists could then be construed as denying one or both of these claims. But how exactly would that go?

The nomic essentialist might be construed as denying the first quidditist claim, as saying that mass and charge and other scientific properties aren't fundamental properties. But such a claim wouldn't on its own imply any distinctive claim about the connection between properties and laws. Also it seems metaphysically questionable, first because it would be unclear what fundamental properties would replace scientific ones, and second because laws involving scientific properties are presumably fundamental facts,[18] and any property occuring in a fundamental fact must surely be a fundamental property.[19]

The nomic essentialist might instead be construed as denying the second quidditist claim, as saying that singular facts about the instantiation of scientific properties—such as the fact that a certain electron e has a certain charge, or that a certain pair of points of space-time p and p' stand in a certain geometric relation—are not fundamental facts, on the grounds that they speak of scientific properties in isolation from the laws. But such facts can't simply be rejected, since they're essential to the description of the world. (The laws aren't all there is to the world; there is also the matter of which of the many histories that are permitted by the laws actually happens!) So new fundamental facts of some sort must be introduced that ground (or replace) these singular facts; and these fundamental facts must somehow vindicate the nomic essentialist idea that properties can't be "understood independently of" laws.

These new fundamental facts might be construed as being relational in some way, so that a complete fundamental fact of x's having a certain mass cannot be formed without somehow bringing in lawhood as well. (Analogy: comparativism about mass implies that my mass isn't "independent" of the masses of other things in the sense that there simply is no fundamental fact about my mass alone.) According to one simple version of this thought, in cases where a quidditist would recognize a fundamental singular fact that a has P, the nomic essentialist instead recognizes a fundamental fact that is partly general:

$$\exists p(\mathscr{L}(p) \wedge a \text{ has } p),$$

where $\mathscr{L}(p)$ specifies P's role in the laws. But now we're back in territory familiar from section 2.1. This claim is incompatible with fundamentality-theoretic analogues of the ground-theoretic principle that existentials are grounded in their instances, such as that if an existentially quantified statement expresses a

---

[18] Caveat: as noted earlier, the literature is dominated by those who make claims about dispositions or causation or powers, not laws, and some of these writers think that facts about laws are not fundamental, but rather emerge from fundamental facts about causation or dispositions or powers.

[19] I call this the principle of "purity"; see Sider (2011, sections 7.2, 7.3, 7.5).

fundamental fact, then so do all of its true instances, or that (more strongly) no existentially quantified facts are fundamental.

Our struggles here and in section 2.1 to articulate nomic essentialism using ground and fundamentality are due to the "hierarchical" demands of those conceptual tools, and evaporate when the issue is formulated with nonhierarchical tools such as modality: 'it's impossible for charge to exist (or things to be charged) without charge obeying certain laws'.[20] But surely, this modal fact ought to follow from some deeper, nonmodal fact about the natures of properties. The modal formulation just hides a genuine problem by considering the issue at a superficial level. Thus we have an instance of our larger theme, that "structuralist" views are more difficult to articulate in postmodal terms.

2.5 Ungrounded or fundamental existentials?

Several of my arguments have rested on principles about existential quantification: that existentials are grounded in their instances,[21] or that the instances of any fundamental existential fact would be fundamental, or that existential facts are never fundamental. But I have said nothing in defence of the principles, other than that they seem plausible. The structuralist might simply deny them; indeed, Dasgupta once floated the idea that structuralism is precisely the denial of such principles (although this isn't his own view[22]).

What can be said in favour of the principles? Their intuitive appeal is undeniable. The suggestion that an object has the property of having charge because it has some property that plays the charge role—where that property is in fact charge—induces a feeling of metaphysical vertigo. But can anything more be said?

One might argue that existentials are analogous to disjunctions, and thus behave analogously with respect to ground. So, since disjunctions are grounded in their true disjuncts, existentials are grounded in their true instances. (An analogous argument may be given for fundamentality.) The argument is somewhat compelling, but the question then becomes how to justify the disjunctions principle.

---

[20] Compare Hawthorne (2001, pp. 369–70) on Russell's argument (recall note 6).

[21] Actually the argument from section 2.1 could get by with the weaker principle that existentials can never be ungrounded, since the nomic essentialist considered there presumably takes $\exists p(\mathscr{L}(p) \wedge$ a has p) to be the ultimate account of a's having P and hence ungrounded. This weaker principle avoids certain objections to the stronger principle; one might be open to strange patterns of grounding at nonfundamental levels while remaining conservative about the most fundamental level. Fine (1994b) discusses a putative case of "reciprocal" essence, in which the essence of Sherlock Holmes is to be assisted by Watson and the essence of Watson is to assist Holmes, and which arguably leads to cyclic grounding: Holmes's existence is grounded in Watson's, and Watson's in Holmes's. Or (to continue with fiction) if the Holmes stories say that someone stole Holmes's boots without specifying any thief in particular, a sort of fictional realist might argue that it's true that someone stole Holmes's boots but that this existential sentence isn't grounded in any instance. Examples like these don't threaten the weaker principle.

[22] Dasgupta (2009, p. 50); Dasgupta (2016a, section 3).

The disjunctions principle and the existentials principle both seem intuitively obvious, but hard to justify on independent grounds.

2.5.1 *The Tractarian and the semi-Tractarian*

One way to defend a claim is to embed it in an attractive general setting. In the present case, the existentials principle (and the disjunctions principle) could be defended by embedding them in the "Tractarian" view that the fundamental facts are all atomic. (Although the Tractarian view doesn't imply any particular story about how existentials (or disjunctions) are grounded, it makes the standard story natural to adopt.)

But although the Tractarian view is intuitively satisfying, it is notoriously hard to uphold. Surely some negations cannot be grounded in fundamental atomic facts, and thus are themselves fundamental facts.

It might be objected that negations can be grounded in fundamental atomic facts after all: an electron's not being 1 g in mass, for instance, is grounded in a fact specifying the mass that it does have.[23] But the idea underlying the objection, namely to ground the absence of one value of a fundamental magnitude in the presence of another, incompatible value of that magnitude, breaks down when we look more carefully at the metaphysics of magnitudes—that is, quantitative properties (Chapter 4). Consider a comparativist view of mass (section 4.3), for instance, according to which the fundamental facts of mass involve fundamental relations of mass-ordering and mass-concatenation; and suppose that the relation of mass-concatenation fails to hold between a certain trio of objects, a, b, and c (that is, a and b's combined masses fail to equal c's). If this negative fact is to be grounded in fundamental atomic facts, those facts must surely be atomic facts about mass-concatenation and mass-ordering; but it's hard to see what facts of this sort would do the trick. In a comparativist metaphysics of mass, there is nothing that stands to a, b, and c's standing in the relation of mass-concatenation in the way that an electron's having an incompatible value of mass stands to its having 1g mass. A mixed absolutist view of mass (section 4.7.3), on the other hand, *does* embrace fundamental, pairwise incompatible values of mass, but a similar argument can nevertheless be given: there seem to be no fundamental atomic facts that could ground the failure of the mixed absolutist's higher-order relation of property-concatenation to hold in a given case.

One might support the existentials principle with a related but weaker "semi-Tractarian" view: that the fundamental facts consist of atomic facts and negations of (what would be) atomic facts. (Again, although this doesn't imply any particular claim about how existentials are grounded, the standard story seems likely once fundamental truth is ruled out.) The semi-Tractarian view faces fewer contrary cases than the Tractarian view (though it is less satisfying). But it too is hard to uphold, for familiar reasons. Where F expresses some fundamental property,

[23] See Armstrong (2004, section 5.2.1) for a critical discussion of this idea.

suppose that everything is F. This fact must have a ground, given the semi-Tractarian view. And the only available ground would seem to be the plurality consisting of all atomic facts Fa. But grounds must necessitate (it's widely assumed), and it would seem that each member of this plurality could hold even if not everything is F: some extra object could have existed and failed to be F.

The argument of the previous paragraph would fail at the last step if everything exists necessarily (Linsky and Zalta, 1994; 1996; Williamson, 1998; 2002; 2013). But even this "necessitism" doesn't answer the underlying concern, only its modal manifestation. The mere fact that it's impossible for there to exist further objects doesn't make the instances of $\forall x F x$ look any more like a (full) *ground* of that fact, since, intuitively, there is no ground-theoretic basis for their being the only instances, only a modal basis. (The necessity of mathematics undermines modal arguments against the absurd view that all mathematical facts are grounded in my existence, but the falsity of that view is nevertheless manifest.) Only a ground-theoretic version of necessitism would really speak to the concern, for instance the view that facts about which objects exist are not "apt for being grounded", in Dasgupta's (2014b) sense. (Being inapt for grounding can be thought of as the ground-theoretic version of Fine's (2005) notion of unworldliness.)[24]

It is sometimes said that $\forall x F x$ is grounded by its instances $Fa, Fb, \ldots$, together with a "totality fact" $\text{Tot}(a, b, \ldots)$ to the effect that $a, b, \ldots$ are the totality of entities.[25] If the totality fact is atomic then semi-Tractarianism would be rescued. But totality facts sure seem like quantificational facts relabeled: $\text{Tot}(a, b, \ldots)$ seems like the fact that every object is either a or b or . . . .

2.5.2 Grounding-qua

Perhaps there is another way out for the semi-Tractarian. Recall a twenty-five-year-old dispute between David Lewis and D. M. Armstrong. In many writings, Armstrong (1980; 1989; 1997) insisted that every truth must have a truthmaker (in arguments for universals, states of affairs, totality facts, and so forth). Lewis objected that this truthmaker principle of Armstrong's was 'an over-reaction to something right and important and under-appreciated' (1992, p. 218), and should be replaced with the principle that truth supervenes on being: two possible worlds with the same individuals and distribution of natural properties and relations over those individuals are alike in every way. What both the truthmaker principle and supervenience-on-being are reactions to, according to Lewis, was an alleged demerit of brute counterfactual or tensed facts, for instance, which would lack a basis in things. The needed "basis in things", Lewis thought, should be cashed out as supervenience on being (the proper reaction) rather than the truthmaker principle (the over-reaction).

---

[24] I myself doubt that any facts are not apt for being grounded (Sider, 2020).

[25] See Armstrong (2004, chapters 5, 6); Fine (2012, p. 62).

But moving beyond that dialectic, there is a parallel issue concerning fundamentality: how to formulate a "completeness principle" (recall section 1.7.2). The fundamental facts must be complete in some sense, to provide a basis for everything else; but in what sense exactly? One idea is that every fact must be grounded in some fundamental facts. This is the position parallel to Armstrong's truthmaker principle, and the one that causes trouble for the semi-Tractarian. But one might seek instead a completeness principle that is analogous to supervenience-on-being, to the effect that differences in facts must be due to differences in fundamental facts. Given an appropriate principle of this sort, the fact that everything is F would not need a ground in atomic facts and their negations; all that would be required is that differences in whether everything is F would need to be due to differences in atomic facts and their negations.

There is an obstacle to this approach. 'Differences in facts must be due to differences in fundamental facts' is vague, and the obvious precisification, namely 'possible worlds that share the same fundamental facts share all facts' is modal, and shares in the failings of other modal attempts to say what should rather be said in fundamentality- or ground-theoretic terms. For instance, it does not meaningfully constrain the relationship between necessary truths and the fundamental facts: for all it says, a Platonist who held that mathematical truths are necessary would be free to refrain from accepting any fundamental basis for them at all.

Nevertheless, one might think, Lewis's approach had a certain payoff that would be nice to obtain by other means. A fundamental account of reality that includes a certain roster of entities or facts shouldn't need a further fact saying that there are no additional individuals or facts.[26] A roster doesn't need to *say* that it is complete, it just needs to *be* complete. The question is how to articulate and secure this, if not by Lewis's approach.

Here is one way that a friend of ground might pursue. In addition to orthodox grounding claims $f_1, f_2, \cdots \Rightarrow g$, in which the grounding facts $f_1, f_2, \ldots$ are said to ground g without regard for what kinds of facts they are, one might invoke a class of "grounding-qua" claims, in which the grounding facts are said to ground qua satisfying a certain condition. Grounding-qua claims must be understood as sui generis, in that they must not be defined as meaning that g is grounded in the orthodox sense by $f_1, f_2, \ldots$ together with the fact that $f_1, f_2, \ldots$ satisfy the condition. The further fact is not part of the ground of g; rather, it is in light of the further fact that $f_1, f_2, \ldots$ ground g.[27]

Grounding-qua claims can be formalized thus:

$$f_1, f_2, \cdots \Rightarrow_R g \qquad (\text{“}f_1, f_2, \ldots \text{ qua standing in } R \text{ ground } g\text{”}),$$

[26] Indeed, the second requirement would lead to awkward results. If the fundamental roster $F_1, \ldots$ must contain a fact to the effect that there are no fundamental facts other than $F_1, \ldots$, then for some i, F_i = the fact that there exist no fundamental facts other $F_1, \ldots, F_i, \ldots$; facts must therefore be in a sense not well founded.

[27] Bader (2019) and Cohen (2020) defend related views.

where $g, f_1, f_2, \ldots$ are facts and R is a relation over facts.[28] The orthodox notion of ground can then be understood as a special case of grounding-qua: $f_1, f_2, \cdots \Rightarrow g$ if and only if $f_1, f_2, \cdots \Rightarrow_{R_\top} g$, where $R_\top$ is the trivial relation that holds amongst any relata whatsoever. Factivity for grounding-qua amounts to this: $f_1, f_2, \cdots \Rightarrow_R g$ only if $f_1, f_2, \ldots$ and g all hold and $f_1, f_2, \ldots$ stand in R. The principle connecting grounding-qua to necessity would be this: $f_1, f_2, \cdots \Rightarrow_R g$ only if $\Box((f_1$ holds $\wedge$ f_2 holds $\ldots \wedge f_1, f_2, \ldots$ stand in $R) \to g$ holds). And the completeness principle for ground should be taken to say that every nonfundamental fact is grounded in some fundamental facts *qua* standing in some relation or other.

Given this setup, the semi-Tractarian view could be defended by claiming that the fact that everything is F is grounded by its instances *qua all and only its instances*. That is, using $[A]$ to denote the fact that A, the claim would be that $[Fa_1], \cdots \Rightarrow_{R_F} [\forall x F x]$, where $a_1, \ldots$ are all the individuals and R_F is the relation that holds amongst some facts if they are all and only the facts consisting of the attribution of F-ness to some individual or other. Unlike the claim that $[\forall x F x]$ is grounded (in the orthodox sense) in its instances, this claim does not have the questionable modal implication that $[\forall x F x]$ holds in every world in which its instances $[Fa_1] \ldots$ hold. Its modal implication (via the connecting principle of the previous paragraph) is merely that it holds in any possible world in which those instances hold and in which they are all and only its instances.

Perhaps even the Tractarian view could be defended. First consider negative facts $[{\sim}Fa]$ where $[Fa]$ is necessarily such that it would be a fundamental fact if it held. In such cases one might claim that $[{\sim}Fa]$ is grounded in the totality of fundamental atomic facts about a, qua that being the totality of fundamental facts about a. This proposal passes the modal test, anyway: where $[F_1 a], \ldots$ are, in fact, all and only the fundamental atomic facts about a, it's impossible for them to continue to be all and *only* the fundamental facts about a while $[{\sim}Fa]$ fails to hold, since then $[Fa]$ would hold, and so would be fundamental by hypothesis, but would be distinct from each of $[F_1 a], \ldots$. So at least in these cases, grounds for negative facts can be found; and perhaps these, together with universally quantified facts (which have already been discussed) suffice to ground all other negative facts.

But there is a real question of the legitimacy of grounding-qua. If the instantiation of some relation by some facts is relevant to those facts' grounding something, then, one might object, the fact that they stand in the relation simply must be counted as part of the ground. The objection certainly carries force in certain cases. It would be absurd to defend the idea that conjunctions are grounded solely in their left conjuncts by saying that $[A \wedge B]$ is grounded-qua-$[B]$'s-holding in $[A]$ alone. $[B]$ must be counted as part of the ground, and cannot be "moved" from that position into the condition on the grounding. More generally, the "import" principle that $f_1, f_2, \cdots \Rightarrow g$ implies $f_2, \cdots \Rightarrow_{\text{being such that } f_1 \text{ holds}} g$ must be rejected. (Even more generally, $f_1, f_2, \cdots \Rightarrow_R g$

---

[28] The order of the arguments on the left is significant, to line them up with R's places).

does not imply $f_2, \cdots \Rightarrow_{\text{being such that } f_1 \text{ holds and } f_1, f_2 \ldots \text{ stand in } R} g$.) What the defender of grounding-qua must say is that although facts that one might have expected to be grounds cannot in general be regarded as mere conditions on ground, in certain special cases they can be, such as the case of the totality condition on the instances of a universal generalization.

2.5.3 *Fundamental concepts and Tractarianism*

We have been viewing these issues through the lens of ground and fact-fundamentality. Let's instead view them through the lens of concept-fundamentality—my own preferred postmodal tool.

In my view, a concept's indispensability for stating simple and strong laws of nature provides a powerful reason to think it is a fundamental concept (recall section 1.8). That is why we regard spatiotemporal concepts, for example, as being fundamental. But quantifiers are needed to state simple and strong laws of nature, and thus should themselves be regarded as fundamental concepts, if we are willing to apply concept-fundamentality to logical concepts (as I am). Thus we should reject the translations of Tractarianism and semi-Tractarianism into the language of concept fundamentality, namely: 'no logical concepts are fundamental concepts', and 'no logical concepts other than negation are fundamental concepts'.

This is not yet an objection to Tractarianism or semi-Tractarianism as originally formulated in section 2.5.1, since those formulations concerned the fundamental facts, not the fundamental concepts. It would be possible to accept fundamental quantificational concepts while agreeing with those original formulations that quantificational facts are never fundamental facts, and in particular that those facts are grounded in their instances (perhaps together with a non-quantificational totality fact, or perhaps qua their being all the instances). But it seems to me that a natural view for a friend of concept-fundamentality is that any fact composed exclusively of fundamental concepts is a fundamental fact, and has no ground.[29] Given that view, even the original formulations of Tractarianism and semi-Tractarianism must be rejected. Moreover, given the view, fundamental existentially quantified facts are unproblematic.

Where does this leave the objections from sections 2.1 and 2.4? The objections that were based on the principle that existential facts are grounded in their instances and the principle that existential facts can never be fundamental can no longer be accepted, since we have rejected those principles. One objection remains standing: the objection from section 2.4 to the fundamentality-theoretic formulation of nomic essentialism that was based on the principle that whenever an existential fact is fundamental, its instances are *also* fundamental. But although this principle is attractive (I myself accept it), we have not yet seen an argument for it. More generally, concerning the view that the fundamental facts concerning the instantiation of scientific properties are all existential in form—call this 'gen-

---

[29] This is the converse of my principle of purity.

eralism about properties'—we have seen no objection other than that it produces metaphysical vertigo.

In our discussion of structuralism about individuals in Chapter 3, we will consider an analogous form of generalism for individuals; and there I will develop an objection that does not rely quite so flat-footedly on metaphysical intuition (section 3.14.1), and which can be applied to generalism about properties (3.17). But for now, the case against the fundamentality- and ground-theoretic formulations of nomic essentialism relies on the intuitive unacceptability of generalism about properties.

2.6 The replacement strategy and resemblance nominalism

In our dealings with problematic entities, it can be liberating to *replace* them with something else altogether, something from which one can recover whatever was of value in our practice of talking about them. Russell (1905), for instance, famously replaced Meinong's ontology of nonexistent entities with his metaphysics of propositional functions and his semantics of descriptions. Replacing problematic entities is somewhat against the spirit of the grounding revolution, which is to retain problematic entities and say that they're grounded. But some cases call for the old-fashioned approach. There simply is no such thing as the golden mountain.

In the case of structuralism, this approach calls for completely eliminating reference to the entities comprising the structure in question—at the fundamental level, anyway—and accounting for the facts in question using new vocabulary that somehow gets at the structural facts directly. There's no guarantee that a suitable replacement theory along these lines exists; part of what's so interesting about structuralism is that replacement theories can be hard to find. But such a replacement theory is desirable if attainable, so it's worth asking whether it's attainable in the case of nomic essentialism.

In fact there is an available replacement theory in the case of nomic essentialism, a theory that is metaphysically tamer than anything mentioned so far. Consider a form of resemblance nominalism which does away with properties altogether, and instead makes use of a primitive plural predicate of individuals, $R(X)$, which may be glossed thus: 'the Xs resemble one another perfectly in some fundamental respect'. (This is just a gloss; no genuine quantification over respects is intended.) Now, resemblance nominalism is usually paired with a reductionist approach to laws of nature, whereas nomic essentialists tend to think of laws as being more robust; but one could add a primitive sentence operator 'it is a law that' to the mix. Thus the fundamental facts, on this view, are given by all the truths that can be expressed in the language of plural quantification plus the predicate R plus 'it is a law that'.

What nomic essentialists really want is for it to just not make sense to talk about permuting properties amongst nomic roles—such permutations are, they think, distinctions without a difference. But the form of resemblance nominalism just mentioned gives them this: permuting makes no sense because there are no properties to permute![30] (Moreover, given this view, a property-theoretic sentence S in ordinary thought will be admissible insofar as its R-theoretic upshots are true. For instance, 'some property P is instantiated by a, b, and c' will be admissible if and only if $R(X)$ for some Xs containing a, b, and c; but given this approach, permutations of properties will not affect the set of admissible property-theoretic sentences.)

I doubt, though, that many nomic essentialists will accept this olive branch.[31]

---

[30] As Jonathan Schaffer pointed out to me, if resemblance nominalism is combined with Lewisian realism about possible worlds and individuals, as it is in Rodriguez-Pereyra (2002), then permutational differences would be allowed after all, since they could be based on relations to merely possible objects.

[31] Here is a related olive branch that will presumably also be rejected: quidditism about spatiotemporal relations and a reduction of all other scientific properties to them as in Hall (2015, section 5.2). Like resemblance nominalism, this view rules out the possibility of permutations of nomic roles (excepting spatiotemporal relations). But see section 4.12.

3

Individuals

Belief in individuals may not be quite universal, but it comes pretty close. We do disagree about cases, as Quine (1948) says. Do there exist particles? Points of space-time? Objects with parts? Holes? Numbers? Propositions? Gods? But nearly everyone accepts some individuals or other, and accepts the concepts we use to think about individuals. We believe, or presuppose, that employing these concepts isn't some colossal metaphysical mistake.

Nevertheless, a number of philosophers have proposed (entertained, fantasized about) rejecting individuals, in one way or another. Even if these proposals are mistaken—as I think they are—studying them, and the intellectual challenges from which they arise, can teach us about the role that individuals play in our theorizing. Until we think about what life without individuals would be like, we won't understand what they are doing for us now.

Moreover, as we'll see, the plausibility of the rejection of individuals is sensitive to the metaphysical tools we use to articulate that view. Rejecting individuals is the sort of structuralist position that is hardest to articulate from a postmodal point of view.

3.1 Entities and individuals

What are "individuals", and what would it mean to reject them? The question is in fact vexed; some clarification and terminological regimentation is in order.

A while ago my daughter started asking the question 'what is a thing?'. Her initial answer was that a thing is something you can touch; then she rejected this on the grounds that atoms are things that can't be touched; later her view was that a thing is anything with a spatial location. But throughout she was clear that not everything is a thing. Feelings clearly aren't things, for instance.

'Not everything is a thing'? Though she was only six years old, my daughter wasn't flatly contradicting herself. She was using 'everything' to express a broad notion of quantification and 'thing' as a restrictive predicate. Had I been a better parent, she would already have known how to express her view in predicate logic: $\sim\forall x\,Tx$. In this chapter we will need a similar distinction, though to avoid appearance of paradox I will use 'individual' for the restrictive predicate rather than 'thing'.

So: we have on one hand the quantifiers 'something' and 'everything', and on the other hand the predicate 'individual'. The quantifiers are to be understood entirely unrestrictedly.[1] If there are physical objects (however large or small), artefacts (such as tables and chairs), social entities (such as governments and economies), locations (times, places), events, purely mental entities (such as Cartesian souls), abstract objects (such as numbers, propositions, or universals), or divine beings, then the quantifiers range over them. (Indeed, that statement is tautological, given the intended lack of restriction on the quantifier at the beginning of the previous sentence—'if *there are* . . .'.)

It will be convenient to have a predicate whose interpretation is broad in a way that mirrors the broad interpretation of the quantifiers. I will use 'entity' to play this role. Thus absolutely everything is an entity, on my usage.

And thus 'entity' must be sharply distinguished from the more restrictive predicate 'individual' (analogous to my daughter's restrictive predicate 'thing'). Not everything, and so not every entity, need be an individual. Individuals are entities that play a certain role that is familiar from ordinary thought and from traditional metaphysics. Other words for roughly the same idea include 'thing', 'object', and 'particular'. Paradigms include the ordinary material objects of common sense as well as the particles of classical physics. Rejecting individuals amounts to denying that any entities play this familiar role.

Let us return to entities, the broader notion. The facts concerning entities—the quantificational facts—constitute "ontological structure" in Jason Turner's sense:

Ontological structure is the sort of structure we could adequately represent with a pegboard and rubber bands. The pegs represent [entities], and the rubber bands represent ways these [entities] are and are interrelated.

(2011, p. 5)

Linguistically, entities are what we signify using terms, whether singular or plural. A singular term, such as 'the president of the United States', refers (on any given occasion of use) to one entity. A plural term, such as 'The justices of the United States Supreme Court', refers to more than one entity. The entities in question are the justices; each justice is an entity. Though, if there exists the *collection*, or *group*, or *set* of justices, then this is itself an entity (though perhaps not an individual, depending on how the latter term is used).

Terms are ubiquitous; entities are deeply embedded in ordinary thought and language. It is unsurprising, then, that the concept of an entity is also deeply embedded in predicate logic. The function of the most basic kind of sentence in predicate logic, an atomic sentence, is to refer to one or more entities using singular terms, $t_1, \ldots, t_n$ (whether fixedly if the terms are names or variably if

---

[1] Or: unrestrictedly as concerning the intended sphere of application; allegedly indefinitely extensible totalities such as the sets will not be at issue. Also: the lack of restriction concerns the range of the quantifiers, not the grammatical type of their variables: these quantifiers are first-order, binding variables that must occupy term position.

the terms are variables), and say something about them using a predicate R: $R(t_1, \ldots, t_n)$.

Entities are also embedded in scientific thought and discourse. This is so both on the surface and also at a foundational level, given the centrality of predicate logic to modern foundational reconstructions of mathematics and science. It's important not to underestimate the extent of this embedding by confusing entities as a general category with particular sorts of entities, such as *particles*. There are real questions about the status of particles in various physical theories, but these do not call into question entities as such, or the underlying conceptual apparatus of quantification, singular terms, and predicates. Perhaps particles should give way to fields, for instance; but the familiar treatments of fields are themselves based on entities, namely the points of the space on which the field is defined, which instantiate properties (or relations) that determine the field values. (The continuing commitment to entities can be especially easy to miss in the case of quantum mechanics, since there is no single agreed-upon quantum theory one can use to illustrate the commitment.)

Let us now return to our initial question: what is it to reject individuals?

A truly radical rejection of individuals would be better described as the rejection of entities and anything like them. It would do away with the entire conceptual scheme of quantification, and moreover would accept nothing structurally similar to it. Reality has nothing like Turner's pegboard structure, on this view; when we think about the world's contents in terms of entities, we make a deep and fundamental mistake.

The problem with this most radical rejection is that it is wholly unclear what is to go in place of the conceptual scheme of quantification. The rejector would need to make an entirely new beginning on the foundations of mathematics and science, not to mention ordinary thought and talk.[2] In fact there are few if any detailed proposals along these lines; almost no one truly and wholly rejects entities in this sense.[3]

Far more common is to back away from some aspect of the orthodox conception of entities, while retaining enough of its structural core to remain descriptively adequate. The point of introducing 'individual' is to have a term—which will remain vague and schematic—for this orthodox conception. We will consider a number of views below which reject, in one way or another, the orthodox or common-sense conception of a large central class of entities, and my blanket term for this is 'rejecting individuals'.

Let us begin by considering some putative reasons for rejecting individuals. These vary in quality, and in the sort of anti-individualist position they support.

---

[2] See section 3.8 below and Sider (2011, section 9.6.2).

[3] Dipert (1997), for instance, is a pleasingly wild manifesto opposing predicate logic as a foundation for metaphysics, but does not provide a clear positive alternative.

Once the reasons are on the table, we will begin to articulate various forms of anti-individualism.

3.2 Traditional metaphysical arguments against individuals

Suppose you believe in the existence of properties, in some sense. What connects an individual to its properties? I suppose the default view is that the individual is wholly distinct from its properties, and that it *instantiates* them. Now, against this it is sometimes said: if an individual is a distinct thing, over and above its properties, then it has no properties—no properties "in itself", it's often added. So it's a mysterious sort of thing—a "bare substratum". This line of thought is often used to support the bundle theory of individuals, according to which an individual is nothing more than a bundle of properties. (In section 3.9 we will discuss the bundle theory in some detail.) In the terminology of the previous section, the bundle theory rejects individuals conceived as wholly distinct from their properties, though it certainly retains entities, namely the properties.

This argument is very weak. Here is Elizabeth Anscombe's withering criticism:

One of the considerations brought forward in erecting this notion (for it is not a strawman, real humans *have* gone in for it) seems so idiotic as to be almost incredible, namely that the substance is the entity that has the properties, and so it itself has not properties.

(1964, p. 71)

Of course individuals have properties—they instantiate them. Instantiating is the sense of 'having' that the view provides.[4]

A better line of argument in favour of the bundle theory is that various metaphysical puzzles are better resolved by a bundle theory than by a substance-attribute metaphysic. For instance, Kris McDaniel (2001) and L. A. Paul (2002) have both argued that the bundle theory best resolves the puzzle of material constitution. On one hand, a statue seems identical to the hunk of matter from which it's made, since they seem to share exactly the same parts. On the other hand, they seem distinct since they apparently have different modal properties: the hunk but not the statue could survive being squashed. McDaniel and Paul's solution is that the statue and hunk do not have the same parts after all. Each is an aggregate of properties, and in the bundles are included different properties. For instance, on Paul's view the statue includes certain modal properties that aren't included in the hunk.[5]

I have no objection to the general form of argument here. But such arguments are highly defeasible. Like anyone offering such arguments, McDaniel and Paul don't claim that competing resolutions of the puzzles are untenable, just that their own solutions are the most attractive, on various grounds. Thus any new

[4] Anscombe goes on to consider things one might mean by 'in itself'. I talk about this issue in Sider (2006a), somewhat in ignorance of past discussions.

[5] According to McDaniel, the properties in the bundles are tropes, whereas according to Paul they are universals. See also Koslicki (2008), who gives a similar solution to the problem of constitution but without assuming the bundle theory.

considerations added to the mix can tip the scales. Below I hope to add such new considerations, by arguing that the bundle theory itself is seriously problematic. Thus one of the competing solutions to the puzzles must be right after all. And in any case, our focus will be on alleged threats to individuals coming from the philosophy of science.

There is a more primordial traditional metaphysical concern about individuals, however, which has direct analogues in the philosophy of science. It is simply the thought mentioned in section 1.3: that there is no genuine difference between possibilities that differ merely over which individuals occupy which qualitative roles. Consider, for instance, a putative possibility that is exactly like actuality in every respect that can be described without using proper names (or similar conceptual devices), but in which Barack Obama and I have exchanged places, so that Obama was born in New Haven in 1967, grew up in Philadelphia, teaches at Rutgers University, and is writing this book, and I was born in Hawaii in 1961 and was the 44th president of the United States. Many have had the thought that there just is no difference between actuality and this "permuted" possibility. They are distinctions without a difference. Or, one might clarify: they are distinctions without a *fundamental* difference: there are never fundamental differences between possibilities that are "merely permutationally distinct". But if there are individuals—or: if facts about particular individuals are fundamental—then it would seem that there are such differences.

Considerations of this sort are often thought to argue merely for the modal doctrine of antihaecceitism: that merely permutationally distinct scenarios cannot both be possible.[6] But if the considerations were cogent, in my view they would call for something stronger: the rejection of individuals altogether, at the fundamental level. We will return to this in section 3.7.

3.3 Metametaphysical argument against individuals

There is a time-honoured tradition of thinking that metaphysics has "jumped the shark" when it asks questions like these: Are there holes? Shadows? Composite material objects? Can two things be located in the same place, or one thing be located in two places?

But what exactly is the concern with these questions?

One is epistemic: we allegedly cannot know the answers to the questions, so the questions are somehow incoherent. This sort of objector faces well-known challenges. The objector must avoid appealing to epistemic principles that lead to scepticism. ('If your view is true, then it would have been possible for me to have the same evidence but be wrong about X; therefore I don't know X.') The objector must avoid giving arguments that apply to scientific inquiry just as much as metaphysical inquiry. The objector must avoid attributing more hubris to metaphysicians than they really have. (Metaphysicians might reasonably only

---

[6] Although see the classic Adams (1979).

claim to be making educated guesses, which in some sense is all any philosopher ever does.)

A second is metaphysical: the competing answers are alleged to not genuinely differ, so the questions are somehow incoherent. The statement 'holes exist in addition to perforated objects' might be said not to make a genuinely different claim about the real world to the statement 'only perforated objects exist'.

Against either concern a direct argument may be opposed, according to which the questions *are* coherent: the questions are stated using the very same vocabulary as certain clearly coherent questions. For instance, the question of whether there are black holes is obviously coherent; but the ontological questions mentioned above have the same form: are there *F*s? Provided we can make '*F*' clear and unambiguous, it's hard to see what's wrong with such questions.

Or take multilocation and colocation.[7] Outsiders hate these topics, but it's actually very hard to avoid thinking that they're genuine. Questions about the existence and number of physical objects and where they are located are, in general, coherent questions. They are *empirical, physical* questions. So how could the question of whether two physical objects are located in the same place, or whether one physical object is located in two places, be incoherent? The question whether 'there is something that is located here, and there is something that is located there' is obviously coherent, so how could the question of multilocation, of whether 'there is something that is located here and (also) located there' be incoherent?

Pushing this further: suppose you think there are such things as individuals *in* space, and that they're different from space itself. Then your picture of a spatially extended individual—an oval, say—is this:

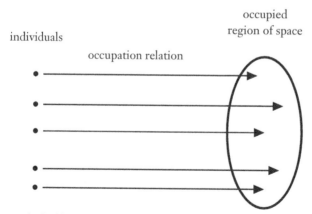

The right half of the diagram represents space itself, and the left half represents the individuals, which are not themselves spatial in the first instance, but only

[7] See, for example, Gilmore (2013); Kleinschmidt (2014); Markosian (1998); McDaniel (2007); Parsons (2007); Saucedo (2011); see Sider (2011, section 5.5) for more on the metametaphysics of the issue.

derivatively spatial, by standing in the relation of occupation to points of space. Given this metaphysics of spatiality, the oval-shaped individual, which is the aggregate of the individuals depicted on the left side of the diagram, is extended because its parts are located at the points in an oval-shaped region of space. But then consider:

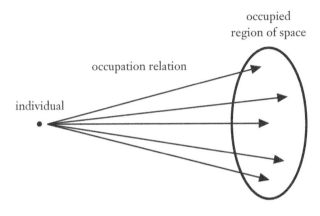

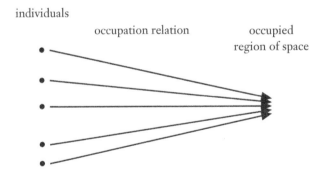

These further pictures are constructed from the same elements as the picture you initially accepted, just arranged in a different pattern, and thus seem themselves to be coherent hypotheses about the makeup of the physical world. The first hypothesis is one of multilocation: there is a single oval-shaped individual with no proper parts, occupying each point in an oval-shaped region; the second hypothesis is that of colocation: many individuals are all located at a single point of space.

In each case—the case of ontological questions and the case of location-al questions—we begin with an initial, ordinary description (an extended oval with many parts; the existence of a black hole); and then the metaphysician uses the very same conceptual materials to construct an extraordinary description (multilocation; the existence of shadows or the nonexistence of tables and chairs); and the philosophical question is whether the extraordinary description is true. If there's a coherent question of whether the ordinary description is true, then, it would seem, there must also be a coherent question of whether the extraordinary description is true.[8]

In the face of this argument, someone who *really* doesn't like the idea that metaphysicians are debating coherent questions—whether for epistemic or metaphysical reasons—might say that the metaphysicians' mistake is in their construal of the initial, ordinary descriptions. For all those ordinary descriptions were taken to involve individuals. In some cases the description concerned individuals of certain types existing, in others it concerned certain patterns of property (or relation) attributions to individuals; but in all cases, individuals were integral to the construal. So maybe individuals are the problem; and maybe the way to resist the argument above that the metaphysical questions *are* coherent is to say that, contrary to appearances, the ordinary descriptions (there is a black hole, an oval is located in a certain region of space) aren't really or fundamentally about individuals after all. Then there would be no way to rearrange the elements of those descriptions to arrive at coherent extraordinary descriptions.

There are serious problems with this sort of motivation. First, it may be that new extraordinary descriptions can be obtained from the vocabulary of the reinterpreted ordinary descriptions. After all, if the "replacement" vocabulary for talk of individuals is to be the basis for a scientifically adequate description of reality, it must surely contain something structurally similar to talk of individuals, in which case new metaphysical questions might well reappear. (The best-worked-out views to be considered below are certainly like this; their supporters are not motivated by the desire to avoid all metaphysical questions.) Second, even if a metaphysics free of individuals is true, the alternative presumably remains coherent, in which case the various competing individuals-theoretic descriptions of reality would seem themselves to be coherent. So it's hard to see how merely upholding the truth of an individuals-free metaphysics would render the metaphysical questions incoherent.

3.4 Structural realists against individuals

Another set of arguments against individuals comes from structural realists, who say that considerations in the philosophy of science, especially the philosophy of physics, call for a new metaphysics that eliminates individuals and replaces them with some sort of structure.[9]

3.4.1 *Rescue from pessimistic metainduction*

Structural realism comes most immediately from an exchange between John Worrall (1989) and James Ladyman (1998).

---

[8] This argument assumes that the coherence of a question is a function of its vocabulary; and one could try denying that assumption by admitting a kind of holism about coherence.

[9] See Ladyman (2014); Saunders and McKenzie (2015) for overviews.

Worrall claimed that to defend against the "pessimistic metainduction", scientific realists should become structural realists. Science has a history of periodic drastic changes. So we should expect, according to the pessimistic metainduction, that few of the central claims of current theories are likely to be true, since those theories will likely be replaced some day by quite different theories—a conclusion that allegedly threatens scientific realism. Worrall's suggestion was then that a scientific realist could resist this argument by appeal to a "structuralist" position. In many cases of drastic scientific change, there is a structural similarity between the new and the superseded theory. So if all that science teaches us is structure, the superseded theory's central claims need not be regarded as inconsistent with those of the new theory.

But Ladyman noted that Worrall's proposal could be read in either an epistemic or ontic sense. *Epistemic* structural realism says merely that all we are justified in believing from science is statements about structure. *Ontic* structural realism—a metaphysical thesis—says that (in some sense) structure is all there is, which in turn is often taken to imply that there are (in some sense) no individuals. Ontic structural realism, which will be our focus, has since been argued for on other grounds, which we will take up shortly.

Our main focus will be on what ontic structural realism (henceforth: 'structural realism') says rather than on arguments in its favour, but it must be said that the argument from the pessimistic metainduction seems weak. First, it's hard to believe that structure (in any helpful sense) really is preserved through radical theory changes in the history of science (though this is hard to evaluate without the sense of sameness of structure being more clearly specified). Second, the pessimistic metainduction just isn't a serious threat to scientific realism if belief is construed as coming in degrees. In light of probable future scientific revolutions, a scientific realist should have a relatively low degree of belief—well below 0.5—in current theory; but that degree of belief could nevertheless be far higher than in all known rival theories. Worrall might reply that this concedes too much to the antirealist since it would undermine the "no miracles" argument for realism, according to which current theories are probably true because otherwise the truth of their predictions about observable matters would be a miracle.[10] But while my realist can't accept exactly that argument (since she thinks that current theories are most likely false), she can in effect accept the no miracles argument for giving current theory the degree of belief that she in fact does. She thinks, let's suppose, that the current theory is 20% likely to be true, that it's 5% likely that some current rival theory is true, that it's 74.9% likely that some product or other of a future scientific revolution (of whose nature she has no inkling) is true; and she reserves a 0.1% probability for a grab-bag of outcomes such as that nature has no regularities at the unobservable level and her theory "just happens" to make true observational predictions. The low degree of belief in the grab-bag reflects her disbelief in

---

[10] See his pp. 110–11 discussion of "conjectural realism".

miracles—her realist faith in the value of explanations citing unobservables, taken at face value.

3.4.2 *Metaphysical undetermination*

Another argument for structural realism is that it allegedly avoids a certain sort of "metaphysical undetermination". In his original paper on structural realism, Ladyman observed that certain physical theories leave open the metaphysical nature of the entities they allegedly concern—in particular, whether those entities count as "individuals" in a certain loaded sense. He mentions two main cases: individuality in quantum mechanics and the debate over substantivalism about space-time. In the former case, there is a question about whether particles in quantum mechanics can be regarded as individuals; and in the latter case, there is a question of whether, as Newton thought, points of space or space-time are individuals, or whether instead, as Leibniz thought—and contemporary advocates of the hole argument think—there are only spatiotemporal relations amongst material objects. In each case, Ladyman claims, structural realism is called for because it can dissolve the question of which of the two viewpoints to adopt:

> Even if we are able to decide on a canonical formulation of our theory, there is a further problem of metaphysical underdetermination . . .. In the case of individuality, it has been shown . . . that electrons may be interpreted either as individuals or as non-individuals. We need to recognise the failure of our best theories to determine even the most fundamental ontological characteristic of the purported entities they feature. It is an *ersatz* form of realism that recommends belief in the existence of entities that have such ambiguous metaphysical status. What is required is a shift to a different ontological basis altogether, one for which questions of individuality simply do not arise. (Ladyman, 1998, 419–20)

The argument here is parallel to that considered in section 3.3: individuals should be rejected because they lead to the existence of certain allegedly re-pellent questions—in this case, according to Ladyman, repellent because they aren't answered by the scientific theories in question.

As with the first argument for structural realism, I'll say only a little by way of critique; our focus is primarily on the position itself. First, as mentioned in section 3.3, any general prohibition of unanswerable questions threatens to overreach. Second (again reprising section 3.3), Ladyman's form of the argument sets us on a quixotic quest for a metaphysical language of science in which one simply can't formulate scientifically unanswerable questions. It seems unlikely that the goal is achievable, given the almost perverse talent we philosophers have for raising devil-ishly difficult questions using any vocabulary with which we're supplied ('How do I know I'm not dreaming?'). Why think that an imagined structuralist reconstrual of physics—if it were ever constructed—would be immune? And it might well come at a price: the theories stated in such a language might score worse on the theoretical virtues. (A "theory" consisting of a mere list of observational conse-quences is perhaps metaphysics-resistant in the desired sense, but is insufficiently explanatory, which is what leads us into the realm of the unobservable in the first place.) The generation of unanswerable questions is the inevitable result of adopting the vocabulary needed to give an explanatory theory of the world.

(Some who are partial to structural realism may wish to refuse the quixotic quest right from the start. Instead of following Ladyman's recommendation to 'shift to a new ontological basis' that does not generate the unanswerable questions, they may instead deny the need for any ontological basis at all. In Chapter 5 I will discuss a view I call 'quotienting', according to which theories can be equivalent even though no underlying account exists of why they are equivalent—no account of the common subject matter of the theories. As we will see in the remainder of this chapter, it is extraordinarily difficult to construct a coherent and attractive structural realist metaphysics. But a quotienter might deny the need for doing so. Instead, when faced with the allegedly problematic cases of metaphysical underdetermination, in which a pair of theories that say different things about individuals are equally supported by scientific evidence, the quotienter can simply assert that the theories are equivalent (rather than attempting to construct an individuals-free metaphysical account of the common reality that the theories are trying to capture). As I will explain in section 5.6.4, this deeply antimetaphysical metametaphysics may indeed appeal to many friends of structural realism. But I do not read structural realists as accepting it. I read them, rather, as presupposing a more traditional metametaphysical outlook and attempting to construct a distinctive structural realist metaphysics. In any case, the discussion of structural realism in this chapter is about that attempt.)

Also, why would the truth of structural realism avoid metaphysical underdetermination? All the individuals-based theories would remain, alongside structural realism, as metaphysical theories consistent with the science in question, and we would still lack purely scientific means to resolve the question of which is correct.[11] Structural realism, conceived as a distinctive metaphysics but supported by the undetermination argument, is dialectically unstable.

In a nutshell, the argument from metaphysical underdetermination is weak because it assumes that metaphysical underdetermination is avoidable and objectionable. A structural realist might concede this, regroup, and offer a "first-order" variant of the argument, claiming that structural realism is the best metaphysical account of the scientific theory in question. For instance, instead of touting structural realism as a way to dissolve the question of substantivalism versus relationalism, one might instead claim that it best solves the problems that these traditional views were responding to. One might, for example, buy the hole argument against substantivalism while rejecting standard relationalism, for one reason or another. Or one might claim that structural realism is the best account of the status of particles in quantum mechanics. (Note, though, that the latter argument needs to establish the superiority of a structural realist account of quantum mechanics over *all* individuals-theoretic accounts, including accounts that

---

[11] See Pooley (2006, p. 91); Saatsi (2010, p. 262).

dispense with particles but retain points of space-time or some high-dimensional space bearing field values.)

On the face of it, the main two arguments for structural realism we have considered differ strikingly in what they support. Only the argument from metaphysical underdetermination (and its first-order variant from the previous paragraph) seems to directly target individuals. The argument from the pessimistic metainduction seems rather to target monadic properties, and recommends replacing them with relations; this leaves open that the relations be orthodox—instantiated by individuals. For the kinds of past theory-changes that undermine our confidence in current theory involve the replacement of properties proposed, or else the replacement of laws governing those properties; but all the theories in question, as usually understood, employ individuals. Thus one would expect the proposed structural replacement of current theory to still be stated in terms of individuals, and to merely replace talk of those individuals' properties with some sort of relational description.

Conversely, the argument from undetermination does not seem to call for replacing properties with relations. What it calls for is understanding the attribution of properties and relations both as somehow not involving individuals.

But perhaps the two arguments may be seen as complementary if the relations of which the structuralist speaks are understood in a certain way. Rather than "concrete" relations, holding (as we usually assume) amongst parts of reality (for instance, the relations that structure space-time), perhaps the relevant relations are more abstract, like relations between entire theories, symmetries, or the relations that structure groups of symmetries. If structural realism is the view that the world somehow consists of structural facts at a high level of abstraction, it might be thought to both secure continuity across theory change and obviate questions about individuals, and thus to be supported by both arguments. However, the considerable unclarity of what structural realism amounts to, which we will discuss in detail in the rest of this chapter, becomes even greater if it is made abstract in this way.[12]

3.5 Dasgupta against individuals

Although I reply to it in section 3.15 below, in my view the strongest argument for a position in the vicinity of structural realism is that of Shamik Dasgupta, according to which we should dispense with individuals for the same reason that absolute velocities should be purged from Newtonian gravitational theory: they are 'physically redundant and empirically undetectable'.[13]

The consensus in the philosophy of physics has been that it's best to eliminate absolute velocities from Newtonian gravitational theory; if we had accepted that

---

[12] See Ney's (2014) review of French (2014), whose structural realism is of this abstract sort.

[13] Dasgupta (2009, (p. 40)). There are precursors of the argument in Horwich (1978, p. 409) and Field (1985, note 15), though Horwich and Field take the argument to be a reductio of its premises.

theory, we should have adopted a conception of the structure of space-time in which there is no intrinsic fact of the matter about which objects are absolutely at rest, and hence no intrinsic facts about absolute velocities, only intrinsic facts about bodies' velocities relative to one another. Instead of combining Newtonian gravitational theory with Newtonian space-time, in which absolute velocities are well defined, it's thought that it should instead be combined with Galilean ("neo-Newtonian") space-time, in which only absolute accelerations, not absolute velocities, are intrinsically well defined.

Why should absolute velocities be purged from Newtonian gravitational theory? Dasgupta mentions two reasons: they're physically redundant, and they're empirically undetectable. Each reason is based on the following observation. Suppose the universe is configured in a certain way at a time t_0. Given the laws of Newtonian gravitational theory, a certain series of configurations at later times results. Now, if the universe at t_0 had instead been given a "velocity boost"—if a certain constant velocity were added to each thing's actual velocity at that time— then the later configurations would have been correspondingly different in terms of absolute velocities. But this is the only difference there would be. For given the laws of Newtonian gravitational theory, adding a velocity boost at t_0 would not affect the masses, or inter-particle distances, or relative velocities, or anything else we can detect, at later times. So, it would seem, we have no way to detect absolute velocities, since any experiment we could perform would yield the same result whether or not the universe initially received the velocity boost. This is the sense in which absolute velocities are empirically undetectable. Moreover, since the masses, inter-particle distances, relative velocities, and so forth evolve in the same way regardless of the initial velocity boost, the absolute velocities of particles seem to be playing no role in the laws of nature. This is the sense in which absolute velocities are physically redundant.

According to Dasgupta, individuals are likewise empirically undetectable and physically redundant. Suppose that the universe is given a permutation rather than a velocity boost at t_0; two individuals exchange their qualitative roles. We choose two individuals, a and b, and in the permuted universe we assign all of b's actual physical states at t_0 (its location, mass, velocity, etc.) to a and all of a's actual states at t_0 to b. Notice that this won't have any effect at all on the *pattern* of masses, inter-particle distances, relative velocities, and so forth at t_0; and given the laws in *any* physical theory, it won't have any effect on such matters at any later time either. The reason is that physical laws are general. The law of gravitation, for example, says that *any* two entities that are separated by a certain distance and have certain masses are subject to a certain gravitational force. It doesn't say: if particular particles Joe and Frank are separated by a certain distance then they'll be subject to a certain gravitational force, whereas certain particular duplicate particles Callie and Iola in duplicate circumstances will not be thus subject. Therefore the initial permutation will result in no differences other than permutational differences at later times, just as the velocity boost affected nothing other than future absolute velocities. Now, we cannot directly observe individuals' identities (just as we can't directly observe absolute velocities).[14] And so, since the initial permutation would have no other observational effects, Dasgupta concludes that individuals are empirically undetectable. Moreover, since the laws generate the same evolution of qualitatively defined states, regardless of the initial permutation, Dasgupta concludes that individuals are physically redundant.

So: according to Dasgupta, the same reasoning that led us to reject absolute velocities in Newtonian gravitational theory should lead us to reject individuals as well—to adopt a fundamental metaphysics in which there are no differences corresponding to a permutation of individuals.

3.6 Mathematical structuralists against individuals

A certain sort of mathematical structuralism can also be viewed as being opposed to individuals, namely the form of mathematical structuralism centred on the slogan 'mathematical objects are just positions in structures'.[15] (Structural realists often view mathematical structuralists as kindred spirits.)

Mathematical structuralism begins with the thought that *all that matters* to mathematics is structure. This thought is brought out by several problems in the philosophy of mathematics.

First, Benacerraf's (1965) famous problem. Arithmetic can be "reduced to set theory" in the sense that we can provide definitions of the key arithmetic concepts under which the truths of arithmetic come out true. There are, for example, the von Neumann definitions of the natural numbers:

$$0 = \varnothing$$
$$s(n) = n \cup \{n\}$$
$$Nn = \forall x((0 \in x \wedge x \text{ is closed under } s) \to n \in x).$$

But other definitions "work" as well; Zermelo, for example, defined $s(n)$ as $\{n\}$. These definitions generate different sequences of "numbers":

$$\varnothing, \{\varnothing\}, \underbrace{\{\varnothing, \{\varnothing\}\}}_{2}, \underbrace{\{\varnothing, \{\varnothing\}, \{\varnothing, \{\varnothing\}\}\}}_{3}, \dots \qquad \text{(von Neumann numbers);}$$
$$\ \ 0 \quad\ 1 \qquad\quad\ 2 \qquad\qquad\qquad\ 3$$

$$\varnothing, \{\varnothing\}, \{\{\varnothing\}\}, \{\{\{\varnothing\}\}\}, \dots \qquad\qquad \text{(Zermelo numbers).}$$
$$\ \ 0 \quad\ 1 \quad\ \ 2 \qquad\ 3$$

The question then, is this. The ability to reduce numbers to sets, one would have thought, shows that numbers are sets. But if numbers are sets, then which set is

---

[14] This argument needs to be handled with care, to avoid the reply: 'we can detect identities after all since we can straightforwardly detect, for example, that *that* is five feet from *that*'. See Dasgupta (2009, pp. 38–47).

[15] See MacBride (2005) for a useful overview.

the number 2? Is it von Neumann's 2: $\{\emptyset, \{\emptyset\}\}$? Is it Zermelo's 2: $\{\{\emptyset\}\}$? It can't be both, since they're distinct from each other. It can't be one rather than the other; that would be arbitrary. So it seems that we must say that it's neither. But no other set-theoretic reduction produces a better candidate to be 2. So 2 isn't any of the candidates; numbers aren't sets after all.

Second, the Caesar problem.[16] The problem, as originally raised by Frege, was particular to his conception of numbers, but it has a more general application. Is the number 3 identical to Julius Caesar? It seems that mathematics doesn't settle the question. Mathematics settles questions like whether $3 = 2 + 1$, because such questions pertain to the structure of the natural numbers. But whether $3 = \text{Caesar}$ isn't about the structure of the natural numbers. Mathematical practice doesn't care whether 3 is identical to or distinct from Caesar.

The third problem is similar: statements of inter-structural identity are not settled by the practice of mathematics (just as statements of identity between mathematical and nonmathematical objects like Caesar are not thus settled). For example, whether the natural number two, $2_{\mathbb{N}}$, is identical to the rational number two, $2_{\mathbb{Q}}$ (more generally, whether the natural numbers are a subset of the rational numbers or distinct from the rationals, albeit isomorphic to a proper subset of them) is again a question that normal mathematics doesn't care about.

Perhaps we may add a fourth problem (although I know of no discussions of it in the literature). Consider a scenario in which the natural numbers are, as a whole, structurally isomorphic to the way they actually are, but in which 2 and 3 have "exchanged places". That is: the entity that we actually denote by '2' bears, in this scenario, the successor-of relation to the entity that we actually denote by '3'; but otherwise everything is the same. One might feel that this putative scenario does not genuinely differ from the actual one—perhaps because of a general rejection of mere permutational differences, or perhaps because of the more specific thought that mathematics does not care about the difference between these scenarios.

The problem in each case is the same. All that arithmetic "cares about" is that it's dealing with an "ω-sequence"—some entities together with a relation ordering them that look like this:

$$\bullet \quad \bullet \quad \bullet \quad \text{etc.}$$

It doesn't matter what those entities are, and it doesn't matter which relation over them is chosen, provided it has the right form.

There are responses to these problems that are broadly structuralist in spirit but don't demand a deeply structuralist metaphysics. For example, one response to Benacerraf, both familiar and conservative, is that ordinary number words are semantically indeterminate over various sets. The von Neumann reduction yields

---

[16] See Frege (1884, sections 56, 66), and MacBride (2003, especially section 6).

one acceptable way of interpreting arithmetical language, the Zermelo reduction yields another, but neither interpretation is determinately correct; it simply isn't semantically settled which of these reductions to sets gives the meaning of arithmetical language. Though it's not mandatory, the view can be combined with the supervaluational account of truth in indeterminate languages, with pleasing results. '2 = the successor of 1' and other normal mathematical statements come out true (since they come out true under any acceptable reduction), but '2 has two members' and '3 = Caesar' come out indeterminate since they're true on some reductions and not on others. Such a view is fine as far as it goes, but it is hard to see how it could be extended to set theory itself. Thus structuralist concerns about set theory are left unaddressed. For all that the conservative response is concerned, set theory is about a distinctive range of individuals and a distinguished relation ∈ over those individuals; there is some unknown answer to the question of whether ∅ is identical to Julius Caesar; there is a distinct scenario in which ∅ and {∅} have exchanged their places in the set-theoretic hierarchy; and so on.

Another response that doesn't demand a deeply structuralist metaphysics is to say that number words don't really function as singular terms, numerical quantifiers aren't really quantifiers, and so on. One version of this is modal structuralism (Hellman, 1989). Here, we paraphrase an arithmetic statement $A(N, 0, s)$ along these lines:

$$\Box \forall X \forall y \forall f ((X, y, f \text{ satisfy the axioms of arithmetic}) \rightarrow A(X, y, f)).$$

But what's relevant for our purposes here—what can be considered a structuralist view about individuals in the current sense—is "non-eliminative" or "ante rem" structuralism, according to which there exist certain distinctive sorts of entities, mathematical structures, and that mathematical objects are "just positions" (in some sense) in these structures. Øystein Linnebo (2008, p. 60) describes the view as follows:

[Mathematical objects] are really just positions in abstract mathematical structures. The natural number 2, for instance, is just the second (or on some approaches, the third) position in the abstract structure instantiated by all systems of objects satisfying the second-order Dedekind–Peano axioms.

Its defenders include Michael Resnik (1981) and Stewart Shapiro (1997).

Our main discussion of the structuralist views themselves will come below, but it's worth pointing out right away that this view doesn't appear to solve any of the problems at all. The view seems to say that there really are these things, structures; there really are such things as *positions* in structures; and these structures and the positions they contain are what mathematics is about. But now consider the third position in the natural-number structure. Is it Julius Caesar? Is it a set? Is it identical to any of the positions in the rational-number structure? These questions seem perfectly well formed, and ought to have answers, if we take the talk of structures and positions in them at face value. Yet the structuralist concerns about the more face-value construal of mathematical objects were precisely that these sorts of questions are illegitimate: they're not answered by standard mathematics, and they shouldn't be regarded as being answered by anything external to standard mathematics.[17]

Relatedly, Platonism is a paradigmatically *non*structuralist view, but there is a sense in which what it says is precisely that numbers and other mathematical entities are positions in structures. Suppose you believe in a *sui generis* set of natural numbers, with a sui generis successor relation. It's natural to describe this as the natural numbers structure, and the individual natural numbers as positions in this structure! So what's distinctive about structuralism; what would it mean to say that numbers and other mathematical entities are "just" positions in structures? These questions need to be answered by a distinctive structuralist metaphysics. (We will return to them in section 3.11.)

3.7 Antihaecceitism

We turn now from the arguments to the views themselves—to an investigation of what it would mean to reject individuals.

Let's begin with a quite conservative view that barely counts at all as rejecting individuals: antihaecceitism. Antihaecceitism is a modal thesis, according to which the nonqualitative globally supervenes on the qualitative—possible worlds that are alike qualitatively are alike simpliciter. Thus there are no possible worlds differing solely over which individuals have which qualitative features. This counts as a rejection of individuals, in the current terminology, only insofar as the embrace of permutationally different possible worlds is counted as part of the orthodox conception of individuals.

Antihaecceitism has been regarded as a good solution to certain problems that are regarded as being modal. Consider, for instance, the debate over substantivalism: the Leibnizian "shift" arguments and the more recent hole argument (Earman and Norton, 1987). In the case of the latter, for instance, the argument is that substantivalism implies indeterminism in general relativity. Choose a bounded region—a "hole"—somewhere in space-time; choose a diffeomorphism over the points of all of space-time that maps each point outside the hole to itself, but (smoothly) maps the points inside the hole to displaced ones inside the hole; and consider a description of reality that's just like actuality except that all features of

---

[17] Also, the second ($3 = $ Caesar?) and third ($2_{\mathbb{N}} = 2_{\mathbb{Q}}$?) problems aren't specific to mathematics. Just as mathematical practice doesn't care whether the number 3 is Julius Caesar, ordinary talk of entities such as *methods* doesn't much care whether, for instance, the optimal method for changing the oil in a car is identical to Julius Caesar. Nor does ordinary talk about methods and *quirks* care whether the optimal method for changing oil is identical to the most common personality quirk amongst philosophers. Our speech about things like methods and quirks imposes certain structural constraints on any candidate entities to be the methods and quirks, but nevertheless allows considerable leeway for what those entities are exactly. One would therefore expect a more general solution, to the second and third problems anyway, than that provided by ante rem structuralism.

matter and space-time—including those of the metric field—that are possessed at a given point in the first description are in the second description possessed by the image of the point under the diffeomorphism. Given the diffeomorphism-invariance of the laws of general relativity, since the first description obeys the laws of general relativity, so does the second description; but since the descriptions are identical at all times before the hole (before some chosen Cauchy surface that's before the hole), determinism fails. Antihaeccitism to the rescue: since the two descriptions are qualitatively alike, they correspond to the same possible world given antihaecceitism, and so the threat to determinism vanishes.[18]

As we saw, postmodalists often regard modal facts as being in a sense epiphenomenal. Given this viewpoint, the antihaecceitist's response to the hole argument, for instance, is unsatisfying. The identification of qualitatively indistinguishable worlds ought to hold because of some fact about the contents of those worlds; it should not be a "bare necessity", in Dasgupta's (2011) terminology.[19]

Moreover, suppose the structuralist is motivated by the intuitive thought that merely permutational differences are not genuine (section 3.2). Much structuralist rhetoric, such as that individuals are just "positions in structures", suggests this thought: how could individuals be mere positions in structures if there are structurally identical scenarios differing over where individuals are located in them? But even if one member of any such pair of scenarios is always impossible, if the scenarios are distinct and moreover pertain to perfectly fundamental matters, the intuitive thought has not been captured: there really is a difference, at the fundamental level, between actuality and a scenario, however impossible, in which Obama and I, or a pair of points of space-time, have exchanged roles. (Analogously, a defender of the hole argument might insist that the sort of determinism that it's important to hold is not defined in terms of possibilities, but rather in terms of fundamental scenarios, whether possible or impossible, and thus is not secured by antihaecceitism.)

One needn't appeal to full-strength antihaecceitism to deny that the descriptions correspond to different possible worlds; one could appeal to a more local modal doctrine. For example, according to Tim Maudlin (1988; 1990), points of space-time possess their metrical features necessarily. This is no more satisfying than antihaecceitism from a postmodal point of view. Relatedly, one might support this modal doctrine with the view that points of space-time have their metrical features *essentially*, in a nonmodal sense of essence.[20] But without some more substantive story about how essential features derive from an object's fundamental nature, this fails to scratch the postmodal itch any better than the merely modal

---

[18] Brighouse (1994); Butterfield (1989); Pooley (2006).

[19] Dasgupta also argues that the modal response leaves the substantivalist without a defence against a certain formidable nonmodal argument, namely his own.

[20] This may be Maudlin's view, particularly in Maudlin (1990). See Teitel (2019) for a detailed discussion of what an essentialist response to the hole argument would need to look like, and for scepticism about whether it improves on a purely modal response.

thesis; recall section 2.3. Relatedly, it would not vindicate (what I take to be) the underlying thought: that there is simply no difference at the fundamental level between scenarios differing solely over which space-time points occupy which geometric roles.

So let us turn to nonmodal and nonessentialist ways of articulating structuralism about individuals.

3.8 Eliminative structural realism

We'll begin with what Stathis Psillos (2001) has called 'eliminative structural realism'. This position results from taking structural realists at face value, as simply rejecting the existence of individuals. Individuals do not exist; only structure exists. In some sense. Here are some representative quotations:

> Robert DiSalle (1994) has suggested that the structure of space-time be accepted as existent without being supervenient on the existence of space-time points. This is a restatement of the position developed by Stein in his famous exchange with Grünbaum, according to which space-time is neither a substance, nor a set of relations between substances, but a structure in its own right . . . . This means taking structure to be primitive and ontologically subsistent.
>
> (Ladyman, 1998, p. 420)

> However, a realist alternative can be constructed. The locus of this metaphysical underdetermination is the notion of an object so one way of avoiding it would be to reconceptualise this notion entirely in structural terms.
>
> (French and Ladyman, 2003, p. 37)

> Ontic Structural Realism (OSR) is the view that the world has an objective modal structure that is ontologically fundamental, in the sense of not supervening on the intrinsic properties of a set of individuals. According to OSR, even the identity and individuality of objects depends on the relational structure of the world. Hence, a first approximation to our metaphysics is: 'There are no things. Structure is all there is.'
>
> (Ladyman and Ross, 2007, p. 130)

These writers by no means speak with one voice, but they appear to share the view that reality is, in some sense, nothing but a qualitative structure, a network of relations with nothing standing in these relations.

Now, the first and most flat-footed objection to this is that relations without relata (or properties without instantiators) are incoherent. We are told to subtract the particular entities from the grid of relations, leaving only the pattern behind, like the Cheshire Cat's smile. As many have pointed out, this would seem to make no sense.[21]

This objection may be perceived as arising from metaphysical conservatism, an unwillingness to "think outside the box" and reimagine metaphysical categories such as structure and relation, or perhaps as blind reliance on "intuition". But either

---

[21] See for instance Greaves (2011); Ney (2014); Pooley (2006); Psillos (2006).

reaction would be profoundly misguided. The complaint is just the insistence that some metaphysical account be clearly specified.

What basic notions is the structural realist proposing? What are the proposed rules governing those notions? And how can those notions then be used in a foundational account of scientific theories?[22] Standard predicate logic is the usual home for talk about relations, and gives clear answers to these questions. You can't just continue as if you accepted this framework—by speaking of relations—but subtract the entities and hope for the best. Entities are too embedded within the standard framework; predicate logic provides no sentences about relations that don't also concern entities. You need to properly specify a replacement framework, some replacement inventory of basic notions, rules governing those notions, and methods for using those notions in foundational contexts.

As Frege and Russell and other pioneers understood, great care is needed to develop the most basic framework for theorizing. Predicate logic isn't just a mindless projection of our conceptual scheme. It was developed, with great labour, in a very unforgiving area, the foundations of mathematics, where errors were bound to (and did) have huge consequences. It took a long time to reach the modern viewpoint.

Sure, it *might* be that this approach involves some illicit projection of our conceptual scheme. Rethinking everything from the ground up—great plan. But it needs to be done with the care of the Fregean tradition in logic: with clear choices made about vocabulary and theory, and a demonstration that the new proposed framework is adequate to the foundations of mathematics and science.

From a postmodal point of view, the demand for clear choices of this sort will be accompanied by a corresponding demand that some particular metaphysics of fundamental reality be given, as articulated with the tools of choice. What is reality ultimately like, according to the eliminative structural realist?[23] The friend of concept-fundamentality, for instance, will ask: if we cannot describe structures, in fundamental terms, as involving the instantiation of relations by objects, then what fundamental concepts can be used to describe structures? No answer whatsoever is given to this question in passages like these:

How—it might be asked—can [structures] be regarded as primary and in some sense prior to [objects], when structures—understood as a system of relations—can only be defined in the first place in terms of objects—the relata? If the structural realist cannot answer this question, then the whole metaphysical project threatens to come undone. . . .

This question forms the kernel of an objection to the ontic form of SR which has been voiced to us by Redhead (in private discussion): If structure is understood in relational terms—as it typically is—then there need to be relata and the latter, it seems, cannot be relational themselves. In other words, the question is, how can you have structure without (non-structural) objects? Here the structuralist finds herself hamstrung by the descriptive inadequacies of modern logic and set theory which retains the classical framework of

---

[22] See Dasgupta (2011, pp. 131–4), who complains about the failure of structural realists to answer such questions, and Dorr (2010b).

[23] Compare French's (2014) discussion of "Chakravartty's Challenge".

individual objects represented by variables and which are the subject of predication or membership respectively. . . . In lieu of a more appropriate framework for structuralist metaphysics, one has to resort to a kind of "spatchcock" approach, treating the logical variables and constants as mere placeholders which allow us to define and describe the relevant relations which bear all the ontological weight. (French and Ladyman, 2003, p. 41)

It's fine to say that a certain vocabulary is second rate, and imperfectly represents the truths that bear all the metaphysical weight. But one must say exactly what those weight-bearing truths are, and exactly how the second rate vocabulary represents them. Similarly, many structural realists emphasize conditions under which theories say the same thing, by structuralist lights, without giving a clear structuralist metaphysics of the shared content of the theories.

Now, a structural realist might simply reject the need to provide a distinctive metaphysics. As I mentioned earlier, in Chapter 5 we will consider "quotienting", which is a radical rejection of the demand to say "what reality is ultimately like"; and structural realists may in the end wish to avail themselves of this "nuclear option". But for now I will continue to take structural realists at face value, as intending to provide a distinctive structural realist metaphysics of fundamental reality. What, then, might that be?

One might have expected structural realists to reject individuals in a truly radical way, by rejecting any conceptual scheme that is anything like that of predicate logic. (Recall section 3.1.) Although some informal remarks suggest such a position, this is just whistling Dixie. No one has even begun to articulate a serious account of fundamental reality along these lines.[24] Instead I'll be considering more conservative attempts.

3.9 Bundle theory

'Relations without relata' (and also the rejection of merely permutational differences) naturally suggests the view that only relations exist. A number of views of this sort are possible, but let's start with the most straightforward: the bundle theory.[25]

Bundle theory is opposed to a traditional dualism of particulars and properties. According to this dualism, particulars are what we typically refer to and quantify over (such as tables and chairs, atoms and planets), and are distinct from and irreducible to the properties that they instantiate. Against this, the bundle theory says that a particular is just a bundle of properties.

---

[24] There are intriguing suggestions by Saunders (e.g. 2016, section 9.1) that the most metaphysically perspicuous description of reality is mathematical, in advance of predicate-logic foundations. (And compare Hall's (2015, section 2) idea that magnitudes rather than properties and relations should be foundational.) I feel the pull, but what exactly is the proposal, and how will it avoid the perceived pitfalls of orthodoxy? I look forward to further development of it.

[25] Dorr (2010b) notes the fit between the bundle theory and structural realism, as well as noting how unpromising the bundle theory is. Some other discussions of the bundle theory: Van Cleve (1985); Hawthorne and Sider (2002); Paul (2002; 2012a; 2017).

Let's understand "bundling" as mereological summation (obeying the usual laws). But not just any mereological sum counts as a bundle: the sum gold + mountain shouldn't count as a bundle since bundles are the replacements for the particulars of ordinary thought and science, and there may not be (as we would ordinarily say) any golden mountains. And even if there are golden mountains, that sum might be disqualified as a bundle because it's incomplete—a gold mountain must have some particular mass and shape.

Many bundle theorists deal with these issues by employing a primitive plural predicate 'compresent'. (The term was originally understood in a spatial sense, as signifying being in exactly the same place, but I do not here intend it that way.) We can then define a bundle as the sum of some maximal plurality of compresent properties (i.e. the sum of some things that are compresent and are not properly among any other compresent things), and we can say that a bundle "possesses a property" if and only if that property is one of a maximal plurality whose sum is the bundle.[26]

Bundle theorists have traditionally divided over whether to identify particulars with bundles of *universals* as in Russell (1940, chapter 6), or with bundles of *tropes* as in Williams (1952). The concept of a universal is the familiar one: if two objects have exactly the same charge, then they share a single putative charge universal. A trope, on the other hand, is a "particularized" property or relation: two objects with the same charge have two numerically distinct charge tropes.

Bundles of tropes behave in many ways like particulars as traditionally conceived. For instance, since tropes can be numerically distinct despite being exactly alike, bundles of tropes can be numerically distinct despite being exactly alike. Many opponents of individuals will, therefore, regard a bundle theory based on tropes as being insufficiently radical. For instance, a bundle theory of tropes wouldn't give Dasgupta what he wants, since any given trope would be unobservable and redundant in his sense: just as permuting the identities of individuals amongst their qualitative roles is a symmetry of the laws, so permuting the identities of distinct duplicate tropes is a symmetry.

However, a bundle theory based on tropes can admit distinct duplicate objects, which has been regarded as a major advantage. If an electron in my pocket and an electron on the table are identified with the bundles of their universals, and if, as one might naturally think, they have the same universals, then they will be wrongly identified with each other. But each can be unproblematically identified with the bundle of its tropes, because the tropes of one are distinct from the tropes of the other.

Bundle theorists who favour universals often appeal to relational universals at this point in the dialectic. The electrons then have different bundles after all: only

---

[26] One might say instead that a bundle "possesses" any property it contains as a part, but that could lead to trouble depending on how part–whole relations between bundles themselves are treated. (See Paul (2002) for discussion.) Since the bundle theorist's treatment of relations in general is problematic, let's set this issue to the side.

the electron in my pocket has the property *being in a pocket* in its bundle. The usual counter is to consider the case where the electrons are alone in the world (compare Black's (1952) spheres), so that not even relational properties will distinguish them. Properties of location at points of substantival space might then be invoked: only one of the electrons contains the property *located at point p*. But on the face of it, this gives up on the bundle theory in the case of points. For it presupposes that point *p* is distinct from other points *q*; but points of space are presumably all intrinsically alike, and we presumably can't include the property *being a point at which electron e is located* in *p*'s bundle if we are including the property *located at point p* in *e*'s bundle.[27]

But there is a more basic objection to the bundle theory that is under the surface in this familiar dialectic. What account is to be given of relations? On the face of it, the bundle theory can't accommodate relations at all, since a relation between two things doesn't "fit" into either of the bundles with which the things are identified.

The need for an account of relations is particularly pressing if the bundle theory is in service of a structural realism based exclusively on relations! But it's of course mandatory for any bundle theory.

The silence of bundle theorists on the matter of relations is striking.[28] At best they tend to speak of relational properties, as we saw above: the property of being in a pocket, the property of being located at point *p*, and so on. But it's really not ok to go on speaking of relational properties without giving any systematic account of this talk. Terms like 'being in a pocket' for relational properties are not proper names, but are rather made up of semantically significant parts (the

---

[27] The universals theorist might deny that the properties of location have the form *located at point p*, and say instead that they are primitive properties (compare Teller (1987)). In effect this would make the properties of location play the role of points of space (or space-time): there would need to be a continuum of the properties (one for each point of space, or space-time); the properties would need to have a geometry and hence would need to stand in spatiotemporal relations to one another, just as points would, etc. Since the locational properties now play the role of particulars, one wonders whether they now raise the same concerns that motived the opponent of individuals in the first place.

[28] There are exceptions. Campbell (1990, chapters 5–6) confronts the problem (in the context of a trope theory) and ultimately rejects relations, embracing a kind of monism (on which see section 3.11). McDaniel (2001) also confronts the problem for a bundle theory of tropes—see later in this section. Paul (2012a, 251–5; 2017, 39–40) argues that relations can be combined into bundles of universals by means of a nonextensional mereology, so that the same parts can have multiple fusions; something about the "intrinsic character" of relations determines the intrinsic structures of these fusions. Perhaps some such account can recover the distinctions I am about to discuss, though it would be good to see the details. (Nonextensional mereology is not enough on its own. Suppose relation *R* and property *F* have multiple fusions, four of which correspond to the predicates '*x* bears *R* to something that instantiates *F*', '*x* instantiates *F* and bears *R* to something', '*x* bears *R* to *y* and *y* has *F*', and '*x* bears *R* to *y* and *x* has *F*'. What determines which fusion attaches to which predicate? A systematic account of the relations' "intrinsic characters" is needed. Paul says that they can be metaphorically understood in terms of relations having "ends" or "places" (compare Fine's (2000) positions). Perhaps taking an ontology of relation-places as the literal truth could play a role in a full account, though it would seem to reintroduce distinctions the structuralist is trying to avoid.)

predicates 'in' and 'pocket'); and this semantic complexity is "load-bearing" in a bundle theory. For one thing, it is essential to giving a systematic reconstruction of ordinary and scientific discourse, to stating (schematic) generalizations like the following: for any predicates R and F, the occurrence in some bundle of the property of *R-ing something that is F* underwrites the correctness of the ordinary or scientific assertion that something Rs something that is F. For another, the complexity of such terms is essential to a systematic account of what configurations of universals are possible, to stating (schematic) generalizations like this: if there exists a property of *bearing R to something that is F* then there must exist a property of *being F*.

Given the role these complex terms play, a full account is needed, in terms the bundle theorist takes to be fundamental, of their function. And on the face of it (although we will consider a contrary viewpoint below), the account will need to treat relations as being prior to relational properties. For this priority is strongly suggested by the fact that terms for relational properties are made up (in a semantically significant way that is essential to their theoretical role) from predicates which apparently denote relations. The relational property *being in a pocket* derives (in some sense) from the relation *being in* (and also the property *being a pocket*).

The bundle theorist, therefore, needs an account of relations—an account of how relations figure in the most basic facts. What might such an account look like? A flat-footed approach would be to include a relation in x's bundle if and only if, as we would ordinarily put it, x bears that relation to something.[29] (Better: comprescence may be applied both to properties and relations; and some properties and relations count as compresent exactly in the scenarios that we would normally describe as containing some particular that has all the properties and bears each of the relations to some particulars or other.) The flat-footed approach is intuitively wrong-headed, and a moment's reflection shows that it's a nonstarter. Consider a situation that involves, as we would normally say, two things, an F and a G, each of which bears relation R to something or other. The flat-footed approach describes the situation as involving two bundles, one containing F, the other containing G, and each containing R; and there is no room to include more information. But the situation remains underspecified. Do the things bear R to themselves? To each other? Subcases resulting from different answers to these questions cannot be distinguished by the proposal, even though they are clearly different structures.

A second approach gives up on a parallel treatment of properties and relations—gives up, that is, on incorporating relations into statements of comprescence.[30] Instead, it identifies particulars with bundles of compresent monadic universals

---

[29] Certain alternatives fare no better: . . . something bears the relation to x; . . . something bears the relation to x or x bears the relation to something.

[30] Bundle theorists who discuss relations tend to just assume this must be given up (see, for instance, Campbell (1990, 98–9)), but see Hawthorne and Sider (2002, 55–67).

only, but then ascribes relations to the bundles as if those bundles were particulars, by saying that the bundles *instantiate* relations. Such a view may seem like an unattractive hybrid, with its different treatment of properties and relations, the former with compresence and the latter with instantiation. But more importantly, it leaves unanswered the problem of duplicate particulars: an electron in my pocket and an electron on the table will be identified with the same bundle of monadic universals and hence with each other.

John Hawthorne (1995) once suggested a solution to the problem of duplicate objects. Concerning Black's spheres, he said: there really is only one sphere, since the sphere is a bundle of monadic universals; but that bundle is multiply located; it is two miles from itself. (In addition to being zero miles from itself—the two are consistent on this view.) After all, Hawthorne pointed out, any believer in ("immanent") universals already accepts that universals can be multiply located. Black's spheres are just more of the same, as are more mundane examples of duplicate objects.

Let's pair this solution to the problem of duplicate objects with the second way for incorporating relations.[31] Thus we identify particulars with bundles of compresent monadic universals, we speak of those universals instantiating relations, and we insist that the impossibility of distinct duplicate particulars is not a problem because their role can be taken over by a single multiply located bundle. Call this the "multilocational" bundle theory.

But as Hawthorne and I (2002) later pointed out, multiply located bundles cannot take over the role of distinct duplicate particulars. The problem is that distinct facts involving the instantiation of relations by bundles cannot be "linked" on this approach. Consider three duplicate particulars arranged on a line, with adjacent particulars separated by one centimetre:

●   ●   ●

Grant for the sake of argument that the multilocationist can account for the geometry of the situation by saying that a certain bundle B is both one and

---

[31] That wasn't quite the way Hawthorne himself intended it. He wrote:

> The bundle theory thus holds that at the metaphysical groundfloor, there are universals standing in relations to each other. Some are clustered together ('compresent' in Russell's lingo), some are at other spatiotemporal relations to each other and to themselves. (p.193)

This suggests utilizing 'instantiation' alone, dispensing with compresence in the present (nonspatial) sense, and defining a bundle as a fusion of a maximal plurality of monadic universals that are spatiotemporally co-located (i.e. at zero distance from one another). This is inadvisable, I think; it would preclude co-location by distinct particles; see Paul (2002, p. 580). McDaniel (2001) takes this approach to a bundle theory of tropes, and there it is less problematic since a single location could contain multiple duplicate tropes. Still, it could not make certain distinctions (which may not, to be sure, be important to make given actual physics), such as that between a location containing a single thing with a certain charge and mass, and a location containing a thing with that charge but no mass at all and a distinct thing with that mass but no charge at all.

two cm from itself.³² But now consider two possibilities based on this setup, each involving the instantiation of a pair of symmetric relations, R and S. In the first case R holds between the left two objects—as we would ordinarily say, presupposing particulars—and S holds between the right two:

In the second case R and S each hold between the left two objects:

³² It's not at all clear that this should be granted: multilocation wreaks havoc on familiar approaches to the foundations of physical geometry. One issue is that of how to account for points of space (or space-time) as bundles of (monadic) properties. There are three options: (i) relationalism: there are no points; (ii) fields: points are bundles of properties such as field values; (iii) there is a distinctive property of *being a point*, and each point is the bundle consisting of this property, so that there exists just one (multilocated) point. Relationalism faces well-known challenges, and the fields option conflicts with a comparativist metaphysics of quantities (see Chapter 4). But the deeper issue is the impact that multilocation would have on axiomatic treatments of the structure of space—treatments that are essential to the numerical representation of space in physics (again see Chapter 4). Consider, for instance, Tarski's (1999) axiomatization of Euclidean space, which is based on primitive predicates of betweenness $Babc$ and equidistance $ab \equiv cd$ over points of space (thus we have bypassed the relationalist option; but similar problems would emerge for relationalist axiomatizations). The points are now bundles; suppose we understand $Babc$ as being true in those cases that we would normally describe as containing some particulars x, y, z, that instantiate bundles a, b, c, respectively, and are such that y is between x and z; and similarly for $\equiv$. Then Tarski's Identity axiom for betweenness, which says that if $Baba$ then $a = b$, fails under the fields option. For some bundle b might be between two of the locations, so to speak, of some distinct bundle a—there might be (as we would ordinarily say) some particular point that instantiates b that is between distinct particular points that each instantiate a:

For the same reason, the Identity axiom for equidistance would also fail. These axioms would not fail under option (iii): distinct a and b cannot be chosen if there exists only one point! But on this option, Tarski's dimensional axioms fail. For instance, his Lower two-dimensional axiom $\exists a \exists b \exists c (\sim Babc \land \sim Bbca \land \sim Bcab)$, which says that there exist three noncollinear points, fails if there is just one point, a, assuming his axiom of Reflexivity for betweenness which implies $Baaa$. The dimensional axioms can also fail given the fields option in certain cases (for instance, one in which, as we would usually say, each point of space has exactly the same field values), as can other axioms, such as Density for betweenness and Uniqueness of triangle construction. Moreover, many of the other axioms don't have their intended import: they can hold or fail for what are, intuitively, the wrong reasons. For instance, the "Transitivity" principle that if $ab \equiv pq$ and $ab \equiv rs$ then $pq \equiv rs$ would fail in this case:

Given multilocation, Tarski's approach would have to be reworked in a drastic way, and it is an open question how this would go. Zooming out: if there is just one point of space, how can different possibilities for the structure of space be characterized? (If a monistic option tempts, see section 3.11.) Fans of multilocation have been too sanguine about its viability as a fundamental theory of physical geometry.

In each case, the facts provided by the multilocationist are the same:

B is one cm from itself

B is two cm from itself

B bears R to itself

B bears S to itself.

Thus the multilocational bundle theory cannot distinguish these two—obviously distinct—possibilities. (Any structuralist about individuals will want to distinguish them; they correspond to clearly distinct relational structures.) What is missing is any way to "link" distinct facts involving the instantiation of relations. The distinction between the two possibilities, intuitively, has to do with the relationship between the third and fourth facts on this list. In the second possibility, there is a "case" of R holding (on the left of the diagram) and S holds *in that very same case*, whereas this is not so in the first possibility.

Particulars (or tropes) would provide the needed links. The third and fourth facts on the list would then be, in the first possibility:

Particular a bears R to particular b

Particular b bears S to particular c,

whereas in the second possibility they would be:

Particular a bears R to particular b

Particular a bears S to particular b.

The linkage between the facts is provided by particulars recurring in them; the different possibilities result from different patterns of linkage. The absence of such linkage dooms the multilocational bundle theory, since it cannot account for the difference between the two possibilities, or myriad other such pairs.

The problem isn't limited to isolated or artificial examples. The multilocationist has no means to represent *any* links between distinct attributions of relations between duplicate particulars. Insofar as our world consists, ultimately, of massive numbers of duplicate points or particulars standing in a network of relations, the collapse of possibilities will be widespread. Nor is the problem merely one of mismatch with "modal intuition". We know perceptually the difference between scenarios with different patterns of linkage. Also the problem would impact the laws of nature. A pair of possibilities might give rise, via the dynamical laws, to distinct outcomes or chance distributions O_1 and O_2, respectively. So if the possibilities are collapsed, the same laws could not hold. At best there could be weaker, disjunctive laws saying that the collapsed possibility gives rise either to outcome O_1 or outcome O_2. If this sort of collapsing is widespread, the "laws" would become so disjunctive and weak as to not deserve the name.

We have been discussing the problems that a universals-based bundle theory has with accommodating relations. Tropes-based bundle theories also face the problem, since relational tropes, just like relational universals, don't seem to "fit" into any one bundle. The move to tropes, however, can help with the second approach to the problem. In McDaniel's (2001) bundle theory, bundles are made of monadic tropes, and relations are then instantiated by these bundles, as in the second approach; but now that the bundles are made of tropes, the problem of possibility-collapse does not arise since there can be duplicate distinct bundles of tropes. Tropes, like particulars, provide the linkage between distinct relational facts.

But McDaniel's approach won't give many of the opponents of individuals what they want, because of the ways in which tropes behave like individuals. We've already noted that particular tropes would be unobservable and redundant in Dasgupta's sense. But further, notice that since McDaniel applies relations to bundles of monadic tropes, his approach won't work at all to eliminate individuals in purely relational structures—that is, networks of individuals where the individuals have no monadic properties at all.[33] (Nor will the multilocational bundle theory.) Thus McDaniel's approach cannot be employed by structural realists who eliminate all properties in favour of relations. Nor can it be employed by more local structuralisms concerning domains of entities that lack intrinsic properties, such as, perhaps, points of space-time or mathematical entities.

3.10 Bare particulars

The slogan 'objects are just positions in a structure' is commonly used to articulate a certain sort of opposition to individuals, especially by structuralists of various sorts, including structural realists and mathematical structuralists. One way to articulate the slogan is this: begin with a qualitative structure—some entities having properties and standing in relations—and now delete the *properties*, leaving only the entities and their relations.

This picture is a somewhat natural fit for structural realists, insofar as their guiding thought is that all we learn from science concerns relations. (It is thus better motivated by the argument from the pessimistic metainduction than the argument from metaphysical underdetermination.) It is also a somewhat natural fit for mathematical structuralists, who emphasize the importance of relations to mathematical practice, and the unimportance of properties.

Early incarnations of Michael Esfeld's "moderate ontic structural realism" were along these lines.[34] Even Ladyman and Don Ross say things that come close, such as the first fifteen words of this statement: 'there are objects in our metaphysics but they have been purged of their intrinsic natures, identity, and

---

[33] McDaniel is explicit about the assumption that each thing has at least one monadic property (pp. 271–2).

[34] Esfeld (2003; 2004). Later incarnations add claims to the effect that objects "have no identity" apart from relations and are "individuated by" them. See section 3.13 below.

individuality, and they are not metaphysically fundamental' (2007, p. 131).[35] (The remaining words will be discussed in section 3.11.) In the case of mathematical structuralism, Linnebo (2008) considers the idea that 'mathematical objects are incomplete in the sense that they have no "internal nature" and no non-structural properties' (p. 61), and quotes Resnik:

In mathematics, I claim, we do not have objects with an "internal" composition arranged in structures, we have only structures. The objects of mathematics ... are structureless points or positions in structures. As positions in structures, they have no identity or features outside a structure.

(Resnik, 1981, p. 530)

The view needs to be refined.[36] It cannot be understood as denying that the entities in question lack properties in a broad sense—Lewis's "abundant" sense (Lewis, 1986a, 59–69)—since any entity would have the property of self-identity, the property of being such that $2 + 2 = 4$, and so forth.

A somewhat better formulation would be that the entities have no *intrinsic* properties. Talk of internal or intrinsic natures is perhaps getting at this formulation. But this is wrong as well if a thing's intrinsic nature—the way it is "in itself"—includes its negative intrinsic nature: the way it isn't, in itself. Even entities that are stripped of "positive" internal natures would then still have intrinsic properties, namely the negations of all the positive intrinsic properties.[37] Furthermore, any entity would presumably have an intrinsic property corresponding to its part–whole structure. This is so even if the entity is a "mereological atom", lacking all proper parts (as Resnick suggests is true of mathematical objects), since the property of being a mereological atom is presumably intrinsic—a matter of what an object is like, considered in itself. Finally, every entity, a, has the property of being identical to a; and such properties are sometimes said to be intrinsic (Eddon, 2011).

In any case, there is a better formulation that sidesteps all these issues about intrinsicality, according to which neither the entities in question nor their parts possess any *fundamental (monadic) properties*; at most, they and their parts instantiate fundamental (polyadic) relations.[38] Call entities of this sort 'bare particulars'.

I have no objections to the truth of this sort of view, in principle anyway. Indeed I suspect that it's true of mathematical entities, given a robust Platonism, and perhaps also true of points of space, or time, or space-time, or other physical spaces. The main question is whether it counts as structuralist in any sense—

---

[35] In a survey of structural realism (2014), Ladyman categorizes a number of views in the vicinity of 'no properties' as versions of structural realism.

[36] The refinements avoid Linnebo's objections.

[37] Compare Bricker (1992).

[38] French (2010, p. 100) makes roughly this suggestion. A mathematical structuralist might add two additional claims: that mathematical entities don't have any proper parts, and that in any particular mathematical structure, the only fundamental relations in which the entities stand are the distinctive relations of that structure—the successor relation, perhaps, for the natural numbers.

whether it gives opponents of individuals what they want. But first I should address the concern that there's something metaphysically objectionable about bare particulars.

Many of the structural realists seem to think there is. Even the advocates of the position regard it as metaphysically daring; one comes across worries about how bare particulars would be "individuated", whether "haecceities" would be needed to individuate them, about what gives them their "identity" or "individuality", and so forth.[39] These concerns are misguided, I think.

One concern in the vicinity presupposes something like the following picture. Reality is fundamentally undifferentiated, not divided into entities. So if you want to say anything about entities, you first have to "individuate" them, by specifying some way in which reality is to be carved up into entities.

But it is entirely unclear what the initial undifferentiated picture is meant to be. It cannot be that of a universe of propertied regions of space or portions of matter,[40] for that universe is a universe of entities: namely, points and/or regions of space, or portions of matter. It's a perfectly coherent picture that in such a universe, there is no *further* privileged carving into the individuals of ordinary thought; indeed, this is the standard view of most "four-dimensionalists" (e.g. Quine (1950); Sider (2001)). But this picture *begins* with an initial description of the universe in entity-theoretic terms. The problem with articulating an undifferentiated, "pre-objectual" picture of the world is in fact very similar to the problem with articulating a coherent structuralism about entities.

And once it's conceded that the fundamental description of reality is to be given in entity-theoretic terms, using the concepts of predicate logic (a natural view to take, given the proven value of such concepts in the foundations of mathematics and science), the concerns about individuation then evaporate. For if the fundamental facts are entity-theoretic then those facts need no further basis. The statement that there exist certain entities standing in certain relations to other entities is, if the relations in question are fundamental, phrased in wholly fundamental terms; why then should the entities' existence need to be grounded in monadic properties?[41]

Perhaps what is thought to need grounding is not the existence of the entities but rather the facts of identity and distinctness between them; perhaps this is where the individuation is supposed to be needed. But once facts about the existence of

[39] See, for example, Esfeld (2003); Chakravartty (2012, p. 197); and also—in a different part of the metaphysics of physics literature, but illustrating the same sensibility—French and Redhead (1988, p. 235). These literatures bear all the marks of an early diet of 1980s UK metaphysics: the prevalence of talk of individuation and individuality, and the fixation on the identity of indiscernibles. In my view this has distorted the discussion. Quine's (e.g. 1948) approach to ontology and identity, coupled with the meta-metaphysics of fundamentality is, I think, a clearer, metaphysically more accurate, and (ironically) metaphysically less loaded basis for the metaphysics of physics.

[40] As in Jubien (1993), say.

[41] I talk more about this in Sider (2006a).

entities are granted to be fundamental, it would be intuitively bizarre to regard facts about their identity and distinctness as needing some more fundamental basis. A domain of entities, intuitively, comes equipped with facts of identity and distinctness; as Quine put it, 'Quantification depends on there being values of variables, same or different absolutely'.[42] For what it's worth, it is hard to conceptualize a fundamental reality with entities but no particular number of them.

But even setting these reservations aside, the denial of fundamental identity and distinctness is off-target. For one thing, it does not meet the felt need for "individuation". For consider Max Black's (1952) duplicate iron spheres, alone in the universe, which are thought to be the kinds of objects needing "individuation", and consider:

> There exist an x and y such that x is an iron sphere and y is an iron sphere and x is two miles from y and x is not two miles from x.

This sentence is true in Black's scenario, and it uses only the non-identity-involving fragment of the language of predicate logic; yet anyone worried about individuation ought surely to worry about what makes it true—about what would individuate the x and y in this scenario (x and y would be nonidentical under any reductionist conception of identity[43]). If the fragment is conceded to express fundamental facts, there is little point in excluding identity. Finally (and most importantly), it will be argued in section 3.13—where we will also have more to say about "individuation"—that this exclusion would not deliver what the opponent of individuals really wants.

The felt need for individuation is thus misguided. The felt need for haecceities is also misguided. Given the view that entity-theoretic concepts (such as quantifiers and names) are fundamental, descriptions of the world such as 'a bears R to b' are perfectly acceptable as descriptions of fundamental reality. There's no need to add "haecceities", fundamental monadic properties A and B possessed by a and b that somehow enable their existence or "individuate" them. Why think that 'a bears R to b and a instantiates A and b instantiates B' is any better than 'a bears R to b'?

The metaphysics of bare particulars is unproblematic. And it would give some structuralists a bit of what they want. For instance, it would give mathematical structuralists the ability to deny that the number zero has any nontrivial intrinsic properties. But it wouldn't give them anywhere near all of what they want. If the successor relation on the natural numbers is a fundamental relation, then exchanging 2 and 3 in its ordering would count as a genuinely different scenario despite being structurally identical to the actual scenario. The question of whether the number 3 is Julius Caesar would have a definite answer, namely *no*, since Caesar is not a bare particular. Inter-structure queries, such as whether 2 is a set and whether $2_{\mathbb{N}} = 2_{\mathbb{Q}}$, would be left intelligible but unanswered, since each of the

---

[42] Quine (1964, p. 101). See also Hawthorne (2003) for a discussion of various parallel issues.

[43] Compare Quine (1961, pp. 325–6)

following three views (among others) is consistent with mathematical entities being bare particulars. (i) There are two nonoverlapping "sui generis" structures, the naturals and the rationals, the former structured by relations $<_N, +_N, \cdot_N$, the latter structured by distinct relations $<_Q, +_Q, \cdot_Q$. (ii) The second of these sui generis structures exists but not the first; talk of the natural numbers is about a substructure of the rationals. (iii) All that fundamentally exists are sets; talk of other mathematical entities is about substructures of the set-theoretic hierarchy. In general, the problem is that, even given the bare particulars approach, there may yet be distinguished entities and distinguished relations that particular branches of mathematics are about.

Nor would bare particulars give Dasgupta what he wants. Of course, his argument doesn't point in the direction of treating particles with mass and charge as bare particulars, but consider his argument as applied to space-time. In a paper on the hole argument (2011), Dasgupta gives arguments against the existence of points of space-time that are similar to his arguments against individuals in general: since permutations of points of space-time are symmetries of the laws, particular points of space are unobservable and redundant. But treating points of space as bare particulars would do nothing to block the conclusion that permutations of points of space correspond to distinct fundamental possibilities.

Would bare particulars give structural realists what they want? Perhaps in regard to the pessimistic metainduction, but not otherwise, as far as I can tell. They don't fit the slogan 'individuals are just positions in structures'. They don't stand in the way of merely permutational differences. They don't block the hole argument.

3.11 Ground and monism

We have been trying to understand what it might mean for individuals to be "just positions in structures". It doesn't mean merely that they lack fundamental monadic properties, as we saw last section. They are to be metaphysically downgraded even more—so much so that permuting them within a structure simply makes no sense. What would this further downgrading be? Two quotations from the previous section are suggestive: Resnik's claim that positions in structures 'have no *identity* or features outside a structure', and Ladyman and Ross's talk of entities being 'purged of their intrinsic natures, *identity, and individuality*' (my emphasis in each case).

Now, it is hard to make any literal sense out of this. (Which is why I ignored the italicized phrases in the last section.) What would it mean to "have identity" "in" or "outside" structures? What would it mean to purge entities of their "identity and individuality"? It might mean purging them of their very existence, but what then would be the entity-free description of fundamental reality? That's exactly the question we've been struggling with in this chapter, so far without success.

But perhaps it means instead: purging them of their *fundamentality*, in that they are somehow *grounded* in structures. In fact, some structural realists and mathematical structuralists have made suggestions along these lines, for instance:

Each mathematical object is a place in a particular structure. There is thus a certain priority in the status of mathematical objects. The structure is prior to the mathematical objects it contains, just as any organization is prior to the offices that constitute it. The natural-number structure is prior to 2, just as 'baseball defense' is prior to 'shortstop' and 'U.S. Government' is prior to 'vice president.'

(Shapiro, 1997, p. 77)

It is thus natural to ask whether various forms of structuralism about individuals should be understood in terms of ontological priority or ground.[44]

One line of thought from section 3.9 converges with this suggestion. There I assumed that talk of relational properties presupposes a prior account of relations. But this might be denied: it might be suggested that, on the contrary, facts about entities standing in relations are to be grounded in properties possessed by larger objects containing those properties as parts. So again, we have arrived at the suggestion that the distinctive structuralist thesis is that the arrow of grounding runs from wholes to parts, rather than from parts to wholes.

An extreme view of this sort is Jonathan Schaffer's (2010) "priority monism", according to which the arrows of grounding (in the concrete realm, anyway) all originate from a single entity, the largest structure of all, the entire cosmos. The sub-cosmic realms of ordinary life and science are posterior to the cosmos itself. Despite its perceived status as high metaphysics *par excellence*, monism is in some ways a natural fit for structural realists, and for other opponents of individuals as well, who all want to prioritize the structure over the nodes in some sense; perhaps this amounts to fundamental reality containing just one entity, a structure, the cosmos, with everything else deriving from it.[45] So I will focus my discussion on monism, although what I will say applies also to less extreme ways to understand structuralism in terms of ground.

My first objection to the monistic conception of structuralism is that it is unsatisfying, for reasons similar to my concerns about the essence-theoretic formulation of nomic essentialism from section 2.3.[46] For now, I continue to understand ground in the way described in section 1.7, as a relation that holds only between facts. (This not how Schaffer construes ground, and so the monism now under discussion is not quite his—see later in this section.) In these terms

---

[44] Fine (1995, p. 270) says that '... holism, in one of its many versions, may be taken to be the doctrine that the parts of a whole can depend upon the whole itself.' See, more recently, French (2014, section 7.6); Linnebo (2008); McKenzie (2014). Cross (2004, chapter 6) takes dispositional essentialism in a similar direction.

[45] They might even prefer what Schaffer calls "existence monism"—a.k.a. Horgan and Potrč's (2000) "blobjectivism"—according to which nothing concrete other than the cosmos exists at all, not even derivatively.

[46] See Sider (2011, sections 8.5–8.6) for more on some of the following issues.

monism says that all facts are grounded in (or else are among) facts about the entire cosmos. But which facts, exactly? What are these fundamental facts about the cosmos, and how do they give rise to everything else? The mere claim that *some* facts about the cosmos ground all other facts is not yet a sufficiently specific thesis about the nature of reality to count as an articulation of the structuralist vision.

This hand should not be overplayed. Unspecific claims of the form 'facts of a certain sort ground all other facts' aren't always objectionable—recall our discussion of physicalism and Wilson's objection to ground (section 1.7.2). Similarly, as we saw in section 2.3, it isn't always objectionable to claim that certain facts are of the essence (in Fine's sense) of a certain entity, without specifying the fundamental facts about the thing in question that mediate this essentialist claim. But sometimes such claims are objectionable, because sometimes a fundamental account is indeed called for. The use of monism to articulate structuralism about individuals is just such a case, since structuralism about individuals is—surely!—meant to be an ultimate account of the nature of reality.

My objection so far has merely been that monistic structuralism is underspecified. But there are also reasons to doubt that it is true, since there are reasons to doubt that an attractive specification *can* be given.

A specification of monism should, at the very least, say something about the ultimate nature of the cosmos. It might, for instance, say something distinctive about which properties and relations are fundamental; or it might say something distinctive about what fundamental facts there are; and in either case it might say something about how sub-cosmic phenomena would emerge.

Here is one such specification: there are no fundamental properties or relations of particles, or of points of substantival spaces of any sort, or of parts of fields, or of any other "sub-cosmic" entities. Rather, there are (i) fundamental properties of the entire cosmos, and (ii) higher-order properties and relations of such properties. The fundamental facts consist of the attributions of such properties and relations, and these facts ground all sub-cosmic facts. Thus what the monist has to work with at the fundamental level is, in effect, a state-space, consisting of properties available to the cosmos, structured by higher-order properties and relations over those properties, and containing a distinguished point, the property that is in fact instantiated by the cosmos.[47]

But such a theory will surely be highly complex. First, a great many fundamental properties will be needed. Each of the completely specific properties that the cosmos as a whole might have must, apparently, be regarded as fundamental. (The pluralist, on the other hand, can recognize a small number of fundamental properties and relations of microscopic entities.) Second, there can apparently be no

---

[47] See Sider (2008a) for more on the idea and the subsequent criticism. Were the higher-order properties and relations eliminated, the view would draw near to the "propositional nihilism" that Turner (2011) formulates and criticizes convincingly, appropriately named the 'bag of facts' metaphysics by Russell (2018).

simple account of the range of possibilities available to the cosmos. (The pluralist can regard this range as being combinatorially generated as all the distributions of the fundamental properties and relations over a vast number of sub-cosmic individuals.) Third, it is doubtful that there will be simple laws, when stated in terms of the properties and relations recognized as fundamental by the monist. At the very least, there is no guarantee that the kinds of simple laws with which we are familiar have simple monist analogues. For the undefined concepts in the simple laws that scientists have actually proposed apply to particles or points or other proper parts of the cosmos. (This is a banal point and should not be controversial: properties about field values at points, for example, are the sorts of properties that a monist of the type currently under discussion cannot recognize as fundamental, since they apply to points of space which are sub-cosmic entities.)

A very different specification of monism would be "even more monistic", and say that there is a single fact, C, that grounds all other facts, namely the fact that the cosmos exists. If this is to differ from the earlier specification, the idea must be that there is no further story to tell about the fundamental nature of the cosmos that explains or mediates its existence grounding all other facts. But this denial makes grounding "magical" in an objectionable way. The monist's picture would be as follows:

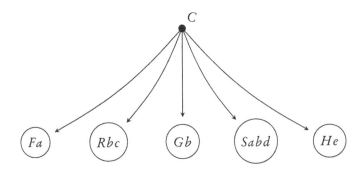

The dot at the top of the diagram represents the fact C that the cosmos exists; the arrows represent grounding, running from the cosmos to the myriad facts of sub-cosmic reality, represented in the diagram by circles containing sentences. What is magical is how the mere fact of the cosmos's existence manages to ground all this immense complexity. The grounding arrows aren't enabled or underwritten by anything fundamental at all.

Suppose someone claimed that fundamental reality consisted of just a single electron, one in my left pocket, say; that one electron gives rise to everything else. This claim is of course absurd, but why? Because, I think, the electron's states are insufficiently rich to give rise to all of the complexities of reality. The arrows of grounding are not "unmediated" or "magical", and cannot be hypothesized to be present in such a complex way unless the entities or facts from which they emerge are themselves correspondingly complex.

The claim that C grounds everything else might seem different from the claim that the electron's states ground everything else. The cosmos is a more complex entity than the electron, after all; it is the most complex entity of all. Relatedly, the monist might replace C, the fact that the cosmos *exists*, with N, the fact that the cosmos has the intrinsic *nature* that it does; and that nature is of course extremely complex. But none of this complexity is recognized at the fundamental level, according to the specification of monism we are now considering. The fact N, for instance, is *not* the fact that *the cosmos contains parts $x, y, z, \dots$ with such-and-such features*, for that would be to mention sub-cosmic features and entities in the fundamental facts and thus to give up on monism. N is rather just the fact that *the cosmos has nature F*, where 'F' is simply a *name* of the intrinsic nature of the cosmos. (One could enrich the conception of F by characterizing its higher-order features in relation to other possible cosmic natures, but that would be a return to the first specification of monism.) In order to bypass the concerns raised above about the monist's fundamental properties and relations, our monist has rejected the demand for an account of the fundamental nature of the cosmos that underwrites all the grounding work that C (or N) must do. The grounding arrows countenanced by this monist, therefore, are no less magical than those that would be needed to ground all facts in the electron in my pocket.

The objection is not that it is impossible for one thing to give rise to many.[48] This is, after all, possible: a lighted match causes both heat and light; a true proposition P grounds the truth of its conjunction with itself, its double-negation, its various disjunctions with other propositions, and so forth. But in these cases of one–many production, the production is mediated by a limited number of general rules which are themselves limited in scope. Combustion doesn't produce just anything: there are laws of nature governing it, specifying its production of heat and light. Moreover, these laws "attach" to the match because of certain of its features. And furthermore they themselves derive from more fundamental laws of nature, which are, we assume, reasonably simple and few in number. Similarly, the grounding of logically complex propositions is presumably governed by a limited number of laws, each with a certain fixed scope (as in Fine (2012)). The problem is that the monist we are considering refuses to recognize any further structure—analogous to the laws of nature and metaphysics in the examples just considered—that mediates its grounding of myriad facts.

In sections 1.7.2 and 1.7.3 a contrast was drawn between a conception of ground as a "super-added force" and a conception on which facts about ground are themselves grounded. A monist might reply that my complaints about magical ground are apt only if the latter conception is presupposed, and say that the grounding facts are unproblematic given the super-added force view.

But holding that these grounding facts are fundamental would amount to giving up on monism. For then, the fundamental facts would include a complete

---

[48] Thanks to Schaffer here for helpful discussion.

specification of the entire world, in sub-cosmic terms! For any sub-cosmic fact F, the monist would be accepting a fundamental fact that C grounds F. (It might be held that only some of these facts are fundamental, but the point remains about those facts.) Moreover, the monist could no longer regard permutations of sub-cosmic individuals over qualitative roles as inarticulable at the fundamental level. If 'C grounds the fact that $\phi(a,b)$' is a statement about fundamental reality, then the permuted statement 'C grounds the fact that $\phi(b,a)$' is as well.[49] Similarly, there would be no obstacle to raising the kinds of questions about identity at the fundamental level that mathematical structuralists think are unintelligible: one could simply ask whether it's a fundamental fact that C grounds the fact that $3 = $ Julius Caesar.[50]

I have so far been understanding ground as a relation between facts. But according to Schaffer's early writings on ground (2009; 2010), grounding relates things of any category, including individuals like the cosmos. So on this view, monism need not be understood as saying that certain *facts* about the cosmos ground everything else, but rather that the cosmos itself grounds everything else. This move to less "structured" relata of the grounding relation heads off some of my critique, for instance my demand for an account of *which* facts about the cosmos ground all other facts. But it runs directly into my critique of the "magical" grounding view, if it is claimed that the grounding of other entities by the cosmos is unmediated by any features of the cosmos.

In fact, however, Schaffer's more recent writings on ground paint a more complex picture. Grounding still relates objects of arbitrary category, but it is mediated by further structure. First, Schaffer now models grounding claims, not as mere binary relations, but rather with structural equations, which specify the systematic dependence of certain "variables" on others (2016). Second, and more importantly, Schaffer's account of metaphysical explanation does not appeal solely to grounding, but also to *laws of metaphysics* (2017a; 2017b). Like laws of nature, laws of metaphysics are general in nature, each one concerning a general sort of input or "source", and specifying, as a function of the particular input, some less fundamental output or "result".

When understood in terms of Schaffer's more structured conception of grounding, monism is *not* like the view based on "magical grounding" above; the emergence of sub-cosmic reality from the cosmos (the production of many from one) would be mediated by variables and laws of metaphysics. But this raises the difficult question of whether my complaints above about underspecificity and complexity can be translated into this framework. Perhaps they can be understood as pertaining to the most fundamental laws of metaphysics that govern the emergence of

---

[49] This last argument could be blocked by a hybrid of magical monism and quantifier generalism (section 3.14) according to which the fundamental grounding statements are all purely general on their "right-hand sides". (Kang (2019) defends a related though distinct view.) The objection to generalism from section 3.14.1 would apply to this view as well.

[50] Also my "purity" principle would seem to be violated (Sider, 2011, sections 7.2, 7.3, 7.5).

sub-cosmic phenomena from the cosmos—call them the fundamental emergence laws—and the kinds of sources involved in these laws. Monism is underspecified (so the first complaint goes) until we have said something about these laws, and in particular about the kinds of sources on their "left-hand sides" (this is the correlate of my earlier demand for an account of cosmos-level fundamental properties and relations). Moreover, the features mentioned in the sources surely cannot be features of proper parts of the cosmos, else the view would not be monistic; but then those features must be many in number since they are not generated combinatorially (this is the correlate of one of my concerns about complexity). Finally, since those features are not generated combinatorially, the fundamental emergence laws seem likely to be exceedingly complex, since it is hard to see how the monist could improve on the list-like "law": 'if the cosmos is like so_1 then the nonfundamental facts are $thus_1$, and if the cosmos is like so_2 then the nonfundamental facts are $thus_2$, and . . .' (this is akin to but not quite the same as the second complaint about complexity above).

It is unclear, though, whether these concerns have force within Schaffer's framework. For one thing, on his view the laws of metaphysics are coarse-grained, individuated by their mapping from sources to results (Schaffer, 2017b, section 4.1). Thus my talk of features being "mentioned" in sources is problematic, and questions of how complexity is to be understood arise. In any case, as I have long suspected, much of the battle over monism must take place at the metametaphysical level, over the tools of metaphysics in terms of which monism is to be understood.

Might there be some ground-theoretic formulation of structuralism that is milder than monism? Linnebo (2008) discusses a formulation of mathematical structuralism according to which the mathematical entities in a structure depend on that structure; and he develops this view using the idea of "abstraction", an idea familiar from the neoFregean programme in the philosophy of mathematics (Wright, 1983; Hale and Wright, 2001). Derivation of entities by abstraction involves (i) an operation O of abstraction, which yields an entity when applied to any member of a certain domain D, (ii) an equivalence relation E on D, and (iii) an abstraction principle specifying the conditions under which O yields identical outputs:

$$O(d) = O(d') \text{ if and only if } E d d' \qquad \text{(for any } d, d' \in D).$$

The entities derived by abstraction need not be identifiable with any entities accepted beforehand; the idea is rather that, given an equivalence relation on a domain of antecedently accepted entities, one can legitimately *introduce* new entities by abstraction. (Much of the controversy about neoFregeanism has centred on what this amounts to and whether it is legitimate.)

Linnebo's key idea is that structures, and positions in structures, are derived by abstraction from set theoretic relations and the members of their fields. The abstraction operation for structures, ‾, yields a structure for any set-theoretic relation (set of 'tuples), and is governed by the abstraction principle that isomorphic relations yield identical structures:

$$\overline{R} = \overline{R'} \text{ if and only if } R \cong R'.$$

Positions in structures, which Linnebo calls *offices*, are obtained by abstraction from "rigid relations"—relations where the only isomorphism between the relation's field and itself is the identity function—and members of their fields via an operation τ, governed by this abstraction principle:

$$\tau(x, R) = \tau(x', R') \text{ if and only if } \exists f(f : R \cong R' \wedge f(x) = x'). \qquad (\tau)$$

(The position in the structure associated with R that corresponds to x is identical to the position in the structure associated with R' that corresponds to x' if and only if x is mapped to x' by some isomorphism between the fields of R and R'.)

Now, on one conception of abstraction (perhaps a more metaphysically loaded conception than the originators of the neoFregean programme had in mind), the objects derived by abstraction exist in virtue of the objects from which they're abstracted. Given this conception, Linnebo's structures and offices depend on sets. Linnebo speaks of there being weak dependence of the offices on the structures, and of the structures on the offices, but really what we have is more like an ontological common cause: there are some sets, R and the x_i, such that *both* the offices $\tau(x_i, R)$ and the structure $\overline{R}$ depend on them.

Now, as Linnebo points out, this view can't be used in a defence of structuralism about absolutely all mathematical entities, since we need some nonstructural mathematical entities to perform abstraction on. Linnebo's argument is that structuralism should be defended for only some mathematical entities, and in particular, not for sets. But even in the case of nonsets, such as the natural numbers, it is not clear that the view delivers what mathematical structuralists want. First, the view implies that the question of whether the natural number $2_{\mathbb{N}}$ is identical to the rational number $2_{\mathbb{Q}}$ has an answer: they are distinct since the naturals are not isomorphic to the rationals. Second, the abstraction principles—principles like (τ)—do not themselves solve the Caesar or Benacerraf problems. If those problems are to be solved, it must be by further added claims about the nature of abstraction. Supposing the natural number $3_{\mathbb{N}}$ to be defined as a certain office $\tau(y, S)$ in some structure $\overline{S}$, the principle (τ) does not rule out that this entity $\tau(y, S)$ is Julius Caesar. (The Caesar problem, after all, is in the first instance a problem for definitions by abstraction.) (τ) only tells us when two terms of the form $\tau(x, R)$ denote the same entity; it tells us nothing about when a term of that form denotes the same thing as a term of another form, such as 'Caesar'. The situation is the same with Benacerraf's problem: (τ) leaves it open whether $\tau(y, S)$ is the Zermelo 3, the von Neumann 3, some other set theoretic 3, or some nonset. (Relatedly: principles like (τ) do not block the existence of a possible world in which 2 and 3 have exchanged their positions in the natural numbers structure, since they leave open that terms $\tau(R,x)$ fail to be rigid designators.) The idea must be that $\tau(y,S)$ just couldn't denote Caesar, or some set, since τ terms denote *abstractions*, and abstractions are entities whose particular features have been "abstracted away", entities whose "identities are nothing beyond the structures they belong to". Thus all the weight of delivering a genuinely structuralist picture rests on these claims about the nature of abstraction, an operation which would otherwise be a metaphysical black box. The claims are cryptic; and when made precise they turn into other claims we have been discussing, and will continue to discuss, in this chapter—for instance the idea that objects in structures lack intrinsic features, as discussed in section 3.10, and the idea that objects in structures are grounded in those structures, as discussed earlier in this section.

Setting aside the details of Linnebo's approach, is there some other interesting form of mathematical structuralism that we could state using the notion of ontological dependence? Most roads lead back to territory we've already covered. Suppose we said simply that the natural number $2_{\mathbb{N}}$ is grounded in the natural numbers $\mathbb{N}$, speaking of grounding as a relation between entities rather than facts, and understanding $\mathbb{N}$ as the fusion of all the natural numbers (or, alternatively, their plurality). This claim is a lot like monism, and leads back to the issues discussed earlier in this section. *How* does $\mathbb{N}$ ground various entities and facts? If the demand for an account of how the grounding works is resisted, grounding would become objectionably magical. And so on. And even if the claim that $2_{\mathbb{N}}$ is grounded in $\mathbb{N}$ were admitted as acceptable metaphysics, it isn't clear that it would give mathematical structuralists what they want. How exactly would it block cross-structure identities (for example)? What will it say about the Caesar problem? (Maybe it just says that $3 \neq$ Caesar since the latter isn't dependent on any structures; but how is that an advance from what the defender of bare particulars would say?) How would it block the existence of duplicate structures? How would it block permutations of entities within structures?

3.12 Indeterminate identity

Mathematical structuralists are moved by the thought that there is something illegitimate about questions of inter-structural identity, questions such as whether $2_{\mathbb{N}} = 2_{\mathbb{Q}}$, or $3 =$ Caesar. Thus ante-rem structuralism might be taken to crucially involve the claim that there is "no fact of the matter" whether such identities hold (as in Resnik (1997)).

The suggestion would not deliver all of what mathematical structuralists want. The existence of a possibility for fundamental reality, distinct from the actual one, in which $2_{\mathbb{N}}$ and $3_{\mathbb{N}}$ have exchanged their places in the natural numbers structure $\mathbb{N}$ would presumably be repugnant to many structuralists, and appears to clash with the slogan 'numbers are just positions in structures'. Yet denying the factuality of inter-structural identity statements does nothing to avoid it. (The denial does not undermine the factuality of the *intra*-structural distinctness of $2_{\mathbb{N}}$

and 3_N, nor the factuality of the distinctness of the imagined possibility from the actual one given the distinctness of 2_N and 3_N.[51]) Still, the denial could perhaps supplement some of the views we've already considered.

A certain bad habit in the philosophy of language has often been remarked on: dumping everything that's ill-understood into the pragmatics bin. The following is also a bad habit, and for similar reasons: if you don't know how to answer a question, just say 'no fact of the matter', and leave it to others to figure out what that really means. It's a bad habit because there are hard and pressing questions about what it really means, especially at the fundamental level; and whether the no-fact-of-the-matter claim is feasible or helpful depends on how those questions are answered.

The clearest understanding of 'no fact of the matter' involves semantic under-specification: there's no fact of the matter whether ϕ if and only if ϕ is true on some sharpenings—ways of making hitherto-unmade semantic decisions—and false on others. On this model a limited structuralism which exempted set theory would certainly be straightforward. Talk of the natural numbers, for example, could be regarded as semantically underspecified, with the sharpenings including various reductions of natural numbers to sets. '$2_N = \{\{\varnothing\}\}$' would be true on a "Zermelo sharpening", false on a von Neumann sharpening, and hence indetermi-nate. But this is just the familiar and conservative reaction to Benacerraf's problem mentioned in section 3.6; we are looking for a more radical alternative.

No-fact-of-the-matter is tempting when talking about fiction. Is the Buffy the Vampire Slayer of the television series identical to the Buffy of the first movie? To the Buffy of the comic books? No fact of the matter, one might want to say. (The temptation isn't limited to identities: does Buffy have a mole between her toes?)

This could be combined with a semantic underspecification approach, as follows. In a fundamental language, it's true to say that there are no fictional characters. But we can provide rules for speaking as if there are:

> Say 'there is a fictional character who does so-and-so' if some fiction contains appropriate sentences entailing 'someone does so-and-so'.
> Say 'fictional character C does so-and-so' if some fiction contains appropri-ate sentences entailing 'C does so-and-so'.

These need refinement along multiple dimensions, but imagine this accomplished, and imagine putting forward sufficiently many such linguistic rules so that in a large range of cases it is clear what the rules require one to say. Perhaps we have then introduced a new language, in which sentences of the form 'there is a fictional character who . . .' are true. ('There is' might mean something different in this new language from what it means in a more austere language.) But now, what of sentences about fictional characters that are not settled by the rules? Suppose that the Buffy canon entails neither 'Buffy had a mole' nor 'Buffy did not have a

---

[51] Distinctness for possibilities is discussed further in section 3.13.

mole'. The rules neither prescribe saying 'fictional character Buffy has a mole' nor saying its negation. Indeed, it would be natural to include a further meta-rule instructing us to say 'there's no fact of the matter whether fictional character Buffy has a mole' in such a case.

We have, then, a semantic account of discourse about fictional characters in a language in which the quantifiers are "fictional", meaning that the truth conditions for existential sentences in this language are given by nonexistential sentences in a distinct metalanguage; and this account has a natural accompanying semantic-underspecification model of 'no fact of the matter'. A kind of fictionalist about mathematics could say similar things. Recall that Hellman recommends replacing $A(N, 0, s)$ with

$$\Box \forall X \forall y \forall f ((X, y, f \text{ satisfy the axioms of arithmetic}) \rightarrow A(X, y, f)). \qquad (*)$$

We could imagine this theory dressed in fictionalist garb. Instead of *replacing* $A(N, 0, s)$ with (*), a language could be introduced in which $A(N, 0, s)$ is *true* if and only if (*) is true in a (more) fundamental language. We could then introduce a 'no fact of the matter' operator into such a language: 'there is no fact of the matter whether A' would count as being true if and only if neither A nor its negation counts as true according to the preceding rule. Thus 'There is no fact of the matter whether $3 = $ Caesar' would be true in this language.

Another model for construing indeterminacy as semantic underspecification comes from quantifier variance à la Hirsch (2011).[52] According to the quantifier variantist, there are many equally good meanings for quantifiers. Which of these do we mean? We pick out one by talking in a certain way; whichever of those meanings fits our talk best is the one we mean. So when we adopt mathematical axioms, we select a meaning for the quantifiers (as well as the mathematical symbols) on which the axioms come out true. But there may be multiple quantifier meanings that agree on the axioms but differ on other sentences, such as '3 = Caesar'. We would be free to adopt a convention as to whether this sentence is to come out true; this would be cutting down further on the candidate quantifier meanings fitting our linguistic stipulations. But before such a conventional decision, there would be semantic indeterminacy in our quantifiers, and hence in our singular terms, and hence in the sentence '3 = Caesar'.

Benacerraf actually says things in this vicinity, in defence of the idea that there's no fact of the matter whether $3 = $ Caesar:

One might conclude that identity is systematically ambiguous, or else one might agree with Frege, that identity is unambiguous, always meaning sameness of object, but that (contra-Frege now) the notion of an *object* varies from theory to theory, category to category—and therefore that his mistake lay in failure to realize this fact.

(Benacerraf, 1965, p. 66)

---

[52] See Sider (2007) for more along these lines.

We've considered two models for there being no fact of the matter whether $3 = $ Caesar, each based on semantic underspecification, and neither of them the familiar conservative one. Each of them is coherent metaphysics (modulo the coherence of quantifier variance anyway). But neither seems particularly (ante-rem) structuralist either, because of the way each downgrades the fundamentality status of mathematical entities.[53] Are there more "inflationary" models of no-fact-of-the-matter?

One such model is a radically plenitudinous form of Platonism, in which, to a first approximation, any consistent theory of abstract entities is true of its own distinctive entities, with distinctive fundamental properties and relations.[54] This would result in there being no fact of the matter in many cases, since there would (or at least might) be no single entities we are referring to with the expressions of any bit of mathematics. (Even if the axioms were categorical, as in second-order Peano arithmetic, there might be duplicate structures.) Again: coherent metaphysics (modulo certain worries about how to avoid inconsistency between the descriptions of different sorts of mathematical entities), but not particularly structuralist.

Another model would invoke worldly indeterminacy. We might have a fundamental nonclassical logic of some sort, or a fundamental notion of determinacy. Speaking for myself, this seems metaphysically extravagant, and not worth what it buys. (Scientific realists don't generally respond to the underdetermination of theory by data in the philosophy of science by saying 'no fact of the matter'.)[55] But it does give some structuralists some of what they want.

This section has focused on mathematical structuralism; but have we opened up any promising new avenues for nonmathematical forms of structuralism? Not really. Indeterminate identity was tempting in the mathematical case because of the Benacceraf problem and the problem of inter-structural identities; but in the nonmathematical case, the denial of merely permutationally distinct possibilities is a more central concern, and as we saw, indeterminate identity is of no help there.[56] But if an attack on identity still tempts, please continue to the next section!

3.13 Weak discernibility and individuation

Some structural realists have emphasized the distinction between "weak" and "strong" discernibility.[57] Entities are strongly discernible, relative to some language, when some formula of that language with one free variable is satisfied by one but not the other; entities are weakly discernible when some formula with two free variables is satisfied by the pair but not by one of them taken twice. For

---

[53] This does not conflict with the aims of Resnik (1997), who disavows such metaphysical ambitions.

[54] Compare Bricker (1992); Eklund (2006)

[55] See also MacBride's critique (2005).

[56] Also there is no nonmathematical analogue of Hellman's approach, or its fictionalist variant.

[57] See Quine (1976); Saunders (2003) on that distinction.

instance, despite failing to be strongly discernible, Black's spheres are weakly discernible: 'x is two miles from y' is satisfied when one sphere is assigned to 'x' and the other to 'y' but not when one sphere is assigned to both variables: the spheres are two miles from each other, but neither is two miles from itself. It is claimed that structuralism demands that objects be "individuated" by their position in the qualitative grid, and that all pairs of distinct objects being weakly discernible, with respect to a language with predicates for qualitative features (but no primitive identity predicate), suffices for this. But the focus on weak discernibility is misplaced, if the goal is a structuralist account of fundamental reality.

The mere claim that every two distinct entities are weakly discernible doesn't *itself* constitute such an account. Even the most committed opponent of structuralism could grant that, as a matter of fact, any two distinct individuals are weakly discernible.

A stronger claim would be modal: necessarily, any distinct entities are weakly discernible. But this would be unsatisfying if nothing else were said: *why* does this modal claim hold?

Also, the modal claim would not deliver what is arguably a *sine qua non* for structuralism: the rejection of merely permutational differences, as in the hole argument, as in Obama and me exchanging places, and as in $2_\mathbb{N}$ exchanging places with $3_\mathbb{N}$. Let's look at this more closely with a toy example, a pair of very simple possible worlds, w_1 and w_2:

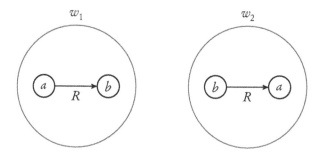

(The only fundamental facts in these worlds are those depicted in the diagram. Thus a and b are the only objects in w_1; and in that world, a bears R to b but nothing else, and thus not to itself. a and b are certain distinct, actually existing entities, and R is a certain actually existing relation.) The modal claim does not rule out w_1 and w_2 being distinct possible worlds. All it demands is that within each world, distinct individuals are weakly discernible. But this holds in both w_1 and w_2. For instance, in w_1, the distinct objects a and b are weakly discernible since a bears R to b but not to itself; and in w_2 those same objects are again weakly discernible since b bears R to a but not to itself.

Nor does the modal claim clear the way for, or otherwise correspond to, any distinctively structuralist metaphysics. Entities that are merely weakly discernible can't be bundles of monadic universals, since they would then be strongly discernible as well. Bundles of relational universals might seem like good candidates for being merely weakly discernible, but this is no good without an account of relational universals. Dasgupta's algebraic generalism, to be discussed in section 3.14, is in essence an account of relational universals, but it does not require weak discernibility. Nor does a metaphysics of bare particulars. No distinctively structuralist approach to fundamental metaphysics is enabled by the claim that distinct objects must be weakly discernible.

After pointing out the distinction between weak and strong discernibility, Simon Saunders goes on to write as if, in a group of entities that are pairwise weakly discernible, there is a sense in which the entities have somehow been reduced to the pattern of relations: 'bodies can be identified by their relations to one another; then a particular body is no more than a particular pattern-position' (Saunders, 2003, p. 163). The second claim doesn't follow from the first. (United States citizens can be identified by their social security numbers; objects can be identified by their unit sets; but in neither case is the former reducible to the latter.) Again: the claim that distinct entities are weakly discernible doesn't itself constitute any particular structuralist reduction of entities, or account of what they are, and nor does it enable any such reduction or account.

Perhaps the claim that entities are "individuated" by their relations is intended to itself constitute a distinctively structuralist metaphysics. This may be how the language of individuation is used in the literature, for instance in the later incarnations of Esfeld's moderate structural realism (Esfeld and Lam, 2008; 2011; Esfeld, 2017; Esfeld and Deckert, 2017). Moderate structural realism consists in part of the claim that objects neither have nor need (fundamental) monadic intrinsic properties, just as the doctrine of bare particulars says. But it goes further, and says that entities are "individuated by" their relations. (It goes even further, by saying, against Saunders, that this individuation requires that distinct entities be strongly, not just weakly, discernible.)

But it is profoundly unclear what "individuation" is supposed to be (which is why I keep scare-quoting that term). Claims about individuation are presumably explanatory claims of some sort; but what facts are explained? Put in terms of grounding: when entities are individuated, what facts about them are grounded?

The facts that are grounded *ought* to be facts about particular entities having properties or standing in relations. Otherwise it would not follow that merely permutational differences are nongenuine—a claim that ought to hold if an entity is "no more than a particular pattern-position". The differences between worlds w_1 and w_2 above consist precisely of differences over facts about particular objects standing in relations. For instance, in w_1 but not in w_2, a bears R to b. If facts such as that a bears R to b are fundamental, not explained or reduced or grounded in any further facts, then w_1 and w_2 are clearly fundamental scenarios, configurations of the fundamental facts. And they are, moreover, clearly distinct scenarios. The distinctness of these fundamental scenarios is anathema to any structuralism worthy of the name; how could any added claim about how entities are "individuated" mitigate this? The resistance needs to come earlier, in the form of a rejection of the fundamentality of facts about particular entities standing in relations.

But how *can* it be facts about particular entities standing in relations that are grounded, when those entities are "individuated" by their relations? Those who bandy such talk certainly don't offer any such account—any account of how facts about particular entities standing in relations are grounded in facts that don't involve entities standing in relations. And this is no accident: the difficulty of finding an entity-free conception of the fundamental facts is precisely what has been driving our discussion in this chapter.[58]

It seems to me, then, that talk of entities being "individuated" by relations is empty. It is meant to rule out the genuineness of merely permutational differences, but it is not associated with any particular metaphysics of entities. It is just words.

But perhaps I have overlooked a possibility for what individuation amounts to? Perhaps it is facts about *identity*, rather than facts about entities standing in relations, that are explained when entities are "individuated". Perhaps, for Saunders, weak discernibility is significant, not because it leads to a distinctive metaphysics of entities, but rather because it leads to a distinctive metaphysics of identity—specifically, that identity can be defined and hence be regarded as nonfundamental.[59] (I don't see how this could be what Esfeld has in mind by individuation, though, since it wouldn't explain his insistence on strong indiscernibility.)

As Quine (1970, p. 63) points out, one can define identity, "or a serviceable facsimile", in a given language by an "exhaustion of combinations" of the atomic predicates of that language. For instance, if the language has just two predicates, a one-place predicate F and a two-place predicate R, one can define a predicate Ixy as meaning $(Fx \leftrightarrow Fy) \wedge \forall z((Rxz \leftrightarrow Ryz) \wedge (Rzx \leftrightarrow Rzy))$. Thus defined I will be identity-like with respect to the language in the sense that it will obey the usual laws: reflexivity and Leibniz's Law with respect to the language. Now, the relation signified by I might not really be the identity relation, if there are distinct objects that cannot be distinguished by the language's predicates. However, suppose that, as it happens, distinct objects are always weakly discernible with respect to the fundamental properties and relations. In that case a definition of I, in a language

---

[58] Esfeld's talk of entities being individuated by relations, together with his insistence on strong—not just weak—discernibility, might suggest the identification of individuals with bundles of relational universals somehow composed of relations; but as we saw in section 3.9, the bundle theory of relations is seriously problematic. Now, as we will see in section 3.14, there are viable views that ban particular entities from the fundamental facts, and indeed rule out permutationally distinct scenarios. But they are nothing like what Esfeld has in mind since they don't even require weak discernibility, let alone strong.

[59] Thanks to Nick Huggett for discussion here.

with predicates for all the fundamental properties and relations, will coincide in extension with identity. For example, if Rxx but not Rxy then the definition above implies that Ixy does not hold (letting z be x, the rightmost biconditional does not hold). Someone might regard the definition as a reduction of identity in such cases.

Now, the reduction is not an independently attractive one. It is intuitively bizarre to reduce identity, as noted in section 3.10. Further, even though the reduction would be "qualitatively parsimonious" since it would eliminate a fundamental concept, it would be "nomically unparsimonious" because it would make the laws more complex. Any law (of nature, logic, or metaphysics) that involves identity would become much more complex, since each occurrence of the identity symbol would be replaced with its extraordinarily long definition.

More importantly, the reduction wouldn't deliver the goods. Recall worlds w_1 and w_2 above. As noted, the merely permutational difference between these worlds is exactly the kind of difference that a structuralist about individuals must reject. But the reduction of identity provides no basis for doing so.

Let's look at this more closely. Worlds w_1 and w_2 are clearly distinct worlds. They differ over which individuals they attribute the relation R to; and the reduction of identity does not make this difference go away. To be sure, the reduction does imply that the distinctness of the worlds is nonfundamental. Just as the distinctness of a and b derives (in either w_1 or w_2) from their qualitative differences (in the sense of Quine's definition), so the distinctness of the possible worlds w_1 and w_2 derives (in actuality and in other worlds as well) from their qualitative differences. For here is a formula with two free variables, w and w', which is (actually and in other worlds) satisfied by w_1 and w_2, but not by w_1 and w_1:[60]

> a bears R to b in w, and b bears R to a in w'.

Moreover, even though the distinctness of the worlds is nonfundamental, the propositions over which they differ—for instance, that a bears R to b—pertain to fundamental matters. Talk of particular individuals like a and b is, on the view in question, perfectly fundamental (it is only the identity relation that is reduced), as is talk of standing in relations like R (we may assume). Thus despite the reduction of identity, there still exist scenarios that (i) are about fundamental reality, (ii) are numerically distinct, and (iii) differ only by a permutation of individuals.

Might it somehow be objected that the reduction of identity applies to transworld identity as well as to intra-world identity, and as a result yields the identification of a in w_1 with b in w_2, and the identification of b in w_1 with a in w_2, making the worlds identical after all? a has, within w_1, exactly the same qualitative

---

[60] I assume that the possible-worlds-theoretic predicate 'in' is contained (or definable) in the language in which identity is defined. Note that although it is allowable to include proper names in the formula, there are also qualitative formulas in two variables that are satisfied by w_1 and w_2 but not by w_1 and w_1, such as 'There exist x and y such that: x bears R to y in w, and y bears R to x in w''.

profile as b has within w_2; so it might be thought that this makes them weakly indiscernible and hence identical; and likewise for b in w_1 and a in w_2.

I don't really see how this argument would go, even on its own terms. But the deeper problem is that the argument is confused from the start, with its pre-Kripkean talk of possible worlds, possible individuals, and "transworld identity". We know from Kripke (1972) (and Plantinga (1976), and others) that possible worlds are *not* like distant planets, existing alongside actuality. Actuality is all that there is; "possible worlds" are just (like everything else) certain actually existing entities—propositions, let's suppose. There are no such objects as "a in w_1" and "b in w_2". There are just a and b, a pair of actually existing entities, and various propositions about them; and these propositions—whether true or false—are themselves actually existing entities. One of these propositions is what we were calling 'world w_1'; it is a proposition saying, among other things, that a bears R to b. Another proposition is what we were calling 'world w_2'; this is a proposition saying, among other things, that b bears R to a. As noted above, these are clearly distinct propositions since they say distinct things about a and b. Of course, a structuralist might attempt to argue that one (or both) of the propositions could not have been true, or that they somehow "correspond to the same possibility"; but these are quite different ways to try to make sense of structuralism, and aren't particularly aided by the reduction of identity.

In the preceding paragraph I simply assumed that Lewis's (1986a) ontology of possible worlds, according to which other possible worlds are "concrete" parallel universes which are ontologically akin to (though spatiotemporally disconnected from) our own universe, is to be rejected. I did not, however, mean to be assuming that *counterpart theory* is to be rejected. Lewis himself first proposed counterpart theory (Lewis, 1968; 1971; 1986a), and presupposed his ontology of possible worlds in doing so, but it is possible to combine counterpart theory with a non-Lewisian reductive theory of possible worlds and individuals (Sider, 2006b). Now, it might be thought that such a combination undermines my argument for the distinctness of w_1 and w_2. If so, my argument might be seen as begging the question against certain structuralists, namely those who regard antihaeccitism as the key to answering certain problems about individuals, such as the hole argument, and who couple their antihaeccitism with counterpart theory.[61]

If the hole problem and the rest were modal in nature, counterpart theory might indeed be the solution. (And note that neither the Quinean reduction of identity nor weak discernibility is needed for that solution; counterpart theory is just a separate approach to these issues.) But we set aside a modal understanding of the problems back in section 3.7. At this point I am assuming that the problems are nonmodal in nature, and that structuralists must offer some distinctive postmodal account of the actual world. The putatively structuralist proposal currently under

---

[61] The relationship between counterpart theory and antihaecceitism is vexed, however; see Skow (2008).

distinction—individuals at the fundamental level, combined with the Quinean reduction of identity—was meant in this spirit. Modality is thus irrelevant if the question is whether the proposal is suitably structuralist. This has perhaps been obscured by the fact that I have been calling w_1 and w_2 *possible worlds*. But I don't in fact mean anything modal here; I am happy to stipulate that w_1 and w_2 are nothing more than propositions about reality at the fundamental level. On the proposal under discussion, propositions about fundamental reality are about particular individuals, and the propositions w_1 and w_2 are, for the reasons given earlier, distinct propositions, counterpart theory notwithstanding.[62] And their distinctness is, I claim, incompatible with the idea that individuals are "just positions in a structure", where that idea is understood postmodally, as a distinctive claim about actuality.

Thus the downgrading of identity does not achieve its desired effect; it does not eliminate merely permutational differences at the fundamental level. Similar remarks apply to other structuralist proposals whose distinctive claim is some sort of downgrading of identity, such as those of Steven French and Décio Krause (2006) and John Stachel (2002). Stachel, for instance, according to Oliver Pooley's interpretation anyway, claims that the facts of numerical distinctness for points of space-time are 'grounded in their standing in the spatio-temporal relations to one another that they do. This in turn is held to prevent our interpreting diffeomorphically related models as representing two situations involving the very same points as occupying different positions in the very same network of spatio-temporal relations' (Pooley, 2006, 104–5). But given the preceding, the second sentence in the quotation does not follow from the first. Eliminating identity from fundamental reality while retaining individuals (like particles or points of space-time) does absolutely nothing to eliminate permutational differences. It is the individuals themselves that must (somehow) be eliminated.

3.14 Algebraic and quantifier generalism

The two views we've discussed that best fit structuralism at an intuitive level, namely the bundle theory and monism, both struggle to provide a sufficiently rich account of the fundamental facts. What is needed is an account rich enough to fully describe physical structures, but not so rich as to allow the distinctions that structuralists eschew: "the structure, and nothing but the structure". In fact there is a view that provides this: Dasgupta's (2009; 2016a) "algebraic generalism".

Like the bundle theorist, Dasgupta accepts an ontology that consists exclusively of universals. But unlike the bundle theorist, Dasgupta makes no use of

---

[62] Talk of propositions might itself be understood counterpart-theoretically, so let me stipulate a non-counterpart-theoretic understanding: propositions here are simply set-theoretic complexes of actual entities such as a, b, and R. It might be objected that my insistence on the importance of identity and distinctness of propositions thus understood is in tension with counterpart theory. If that is right, it strikes me as a real and deep problem with combining counterpart theory—as Lewis for instance seems to do—with antireductionism about individuals.

compresence. Instead, he provides a systematic way for combining universals to form complex universals. As we'll see, his approach incorporates relations and also provides the "linkage" that the multilocational bundle theory could not.

The main trick is to employ an analogue of the predicate-functor approach to first-order logic that Quine brought to our attention in 'Variables Explained Away'. The idea of this approach can be brought out as follows. Simple existential quantifications of one-place predications, such as $\exists x F x$, do not really require the variable x; one could just write $\exists F$, or rather, to use Quine's notation, Der F. Grammatically, Der (for 'derelativization') is a "predicate functor": it attaches to a predicate and forms a complex predicate. When attached to the one-place predicate F, it forms a zero-place predicate—that is, a sentence: Der F.

We could take the negation sign to be a predicate functor as well, rather than a sentential connective. We could then write Der $\sim F$ instead of $\exists x \sim F x$. The predicate functor $\sim$ attaches to the one-place predicate F to form the one-place predicate $\sim F$; then Der attaches to this complex predicate to form the zero-place predicate Der $\sim F$. The other propositional connectives can likewise be treated as predicate functors.

Der doesn't only combine with one-place predicates. In general, it combines with an $n+1$-place predicate to form an n-place predicate, which is—intuitively—the existential quantification of the first place of the original predicate. Thus if R is a two-place predicate, Der R is a one-place predicate meaning 'being R-ed by something', and Der Der R is a zero-place predicate meaning 'something Rs something'.

If a few more functors are added (allowing permutation and addition of argument places), it turns out that the resulting language, "predicate functorese", is equivalent to the usual language of first-order predicate logic in the following sense: for any sentence of first-order predicate logic that contains no names, there is a sentence of predicate functorese that is true in exactly the same models. (The models here are the standard sort, with domains and interpretation functions; the definition of truth in models for sentences of predicate functorese is straightforward.)

Algebraic generalism is based on a language that is like, though not identical to, predicate functorese. In place of predicate functors—expressions that turn predicates into predicates—Dasgupta has term functors, expressions that turn terms into terms. The terms in question name universals, which can have any fixed finite number of places. Given a set of primitive terms for universals, the term functors can then be used to generate terms for complex universals, whose existence Dasgupta also accepts. For instance, a name of a universal can be combined with the term functor c (the analogue of Der), resulting in a complex term referring to a universal that is the existential quantification of the first place of the first universal. If L is the two-place universal (i.e. relation) of loving, then cL is the one-place universal (i.e. property) of being loved by someone, and ccL is the zero-place universal (i.e. proposition or state of affairs) of someone loving someone. Dasgupta also adds a fundamental predicate 'obtains', which he applies to terms for zero-place universals. Thus instead of saying that someone loves someone, he says that ccL obtains. More generally, any name-free sentence of predicate logic has a certain "translation" into term functorese, in which the predicates are converted to names of universals and the entire translated sentence says that a certain complex zero-place universal obtains. 'ccL obtains' is the translation in this sense of '$\exists x \exists y L x y$', although note the abuse of notation (in which I will persist): 'L' in the original is a predicate, whereas in the translation it is a name of a universal.

Term functors provide the "linkage" that was lacking in the multilocational bundle theory. Taking a new example: when two nonsymmetric relations hold between a pair of indistinguishable particulars, the multilocational bundle theory cannot distinguish the relations holding in the same direction from the relations holding in opposite directions:

In each case, the multilocational bundle theorist has only the conjunction of the sentences 'Bundle B bears R to itself' and 'Bundle B bears S to itself'. Dasgupta, though, can distinguish the cases. For the first he would offer the translation of $\exists x \exists y (Bx \wedge By \wedge Rxy \wedge Sxy)$ into term functorese: '$cc(\sigma pB \, \& \, pB \, \& \, R \, \& \, S)$ obtains'; and for the second he would offer the translation of $\exists x \exists y (Bx \wedge By \wedge Rxy \wedge Syx)$: '$cc(\sigma pB \, \& \, pB \, \& \, R \, \& \, \sigma S)$ obtains', where σ, p, and $\&$ are further term functors.[63]

Think of it this way. For the multilocational bundle theorist, an attribution of a binary relation always takes the form 'Bundle B_1 bears relation R to bundle B_2', which is true in cases we'd normally describe thus: $\exists x \exists y (B_1 x \wedge B_2 y \wedge Rxy)$. So in a pair of such attributions:

$$\exists x \exists y (B_1 x \wedge B_2 y \wedge Rxy)$$
$$\exists x \exists y (C_1 x \wedge C_2 y \wedge Sxy)$$

no "link" can be made between the statements; the variables x and y in the second sentence are not bound to the quantifiers in the first sentence. Individuals would provide the needed linkage: names of individuals can recur in distinct attributions of relations, such as Rab and Sba. But Dasgupta can achieve the linkage without names of individuals, since arbitrary sentences of predicate logic without names, including sentences like $\exists x \exists y (Bx \wedge By \wedge Rxy \wedge Sxy)$ and $\exists x \exists y (Bx \wedge By \wedge Rxy \wedge$

---

[63] $\&$ is the term functor for conjunction. σ is a term functor that rotates the argument places of a relation; thus σS is the converse of S. p is a term functor that "pads" a relation by adding a vacuous argument place on the left. Thus pB is a two-place relation that, as we'd normally say, holds between x and y if and only if y has the property B. Argument places were added to B to "line it up" with the two place relations with which it is conjoined. See the appendix of Dasgupta (2009) for more details.

Syx), which attribute multiple relations "at a time", can be translated into term functorese. Only a limited range of sentences of predicate logic are available to the multilocational bundle theorist, namely those sentences corresponding to the attribution of a single relation to bundles. These sentences all take the form:

$$\exists x_1 \exists x_2 \ldots \exists x_n (B_1 x_1 \wedge B_2 x_2 \wedge \cdots B x_n \wedge R x_1 x_2 \ldots x_n)$$

(where B_i is a conjunction $F_1^i \wedge F_2^i \cdots \wedge F_m^i$ corresponding to the m compresent properties in the ith bundle).

Note too that, unlike all the forms of the bundle theory considered above (whether based on tropes or universals), Dasgupta's approach does not require monadic properties, and is therefore friendlier to certain structuralisms. Any sentence of predicate logic without names can be translated into term functorese, and this includes sentences without monadic predicates such as $\exists x \exists y R x y$, which goes into Dasgupta's 'ccR obtains'.

Dasgupta's account is detailed in a way that is important to our discussion in a couple ways. First, in section 3.9 I complained about the bundle theorist's undisciplined appeal to relational properties. Dasgupta also accepts relational properties, but his algebraic generalism specifies clear rules for talking about them. Given an initial stock of fundamental universals, the relational properties—or rather, more generally, complex universals—are then given by the totality of names constructable using the term functors from names for the universals in the initial stock.

Second, I also complained about the structural realist's failure to give a systematic theory. We need, I said, a theory that makes clear choices about its basic concepts and the rules governing those concepts, and which demonstrably is adequate to the foundations of scientific theories. Dasgupta's account clearly satisfies the former: its basic concepts are the names for fundamental universals, the term functors, and 'obtains', and one can write down rules governing those concepts by analogy to the standard inference rules of predicate logic.[64] As for the latter, given the parallelism to predicate functorese, term functorese in a sense has the expressive power of the name-free fragment of standard predicate logic: to every sentence of predicate logic there is a corresponding sentence of term functorese that is "equivalent" in the sense of being true in the same models. This makes possible a systematic reconstruction of a great many foundational theories.

Notice, finally, that algebraic generalism does indeed deliver one of the main things that structuralists want: that there are no fundamental scenarios differing solely by a permutation of individuals over qualitative roles. For at the fundamental level, according to algebraic generalism, there simply are no individuals to permute. Recall the simple worlds w_1 and w_2 from section 3.13:

---

[64] See for instance Bacon (1985).

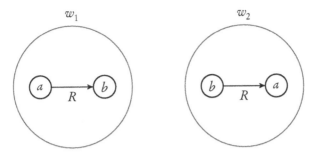

For the algebraic generalist, these diagrams both correspond to the very same fundamental description, namely the translation into term functorese of:[65]

$$\exists x \exists y (Rxy \wedge {\sim}Ryx \wedge {\sim}Rxx \wedge {\sim}Ryy \wedge \forall z(z = x \vee z = y)).$$

Unlike the other views we've considered so far, then, Dasgupta's algebraic generalism both is genuinely structuralist and also has a fighting chance of being the correct metaphysics of the world. That's not to say that it *is* correct; soon we'll be turning to objections. But first it's worth mentioning a closely related view, which Dasgupta (2016a) calls 'quantifier generalism', according to which all fundamental facts can be expressed in the name-free fragment of predicate logic.[66] Thus the fundamental facts might include that every electron has negative charge, that something is at least as massive as something, and so forth.

There is a clear correspondence between quantifier generalism and algebraic generalism. For not only can arbitrary purely general sentences of predicate logic be translated into term functorese (as noted above), the reverse is true as well: there is an obvious method for translating an arbitrary sentence of term functorese into the name-free fragment of predicate logic. Because of this correspondence, quantifier generalism shares many of the virtues of algebraic generalism. It too can account for the linkage missing in the bundle theory; it too allows purely relational structures; it too makes clear choices about the structure of fundamental facts; and so forth. Now, quantifier generalism might seem like a strange fit for certain forms of structuralism, especially structural realism. For there are individuals, according to quantifier generalism; there are fundamental facts that are existentially quantified in form, saying, as it might be, that *there exist* an x and y such that x is as least as massive as y. Nevertheless the view does count as structuralist in a clear sense. By denying the existence of fundamental singular facts—fundamental facts about *particular* individuals, expressible using proper names—the quantifier generalist denies that there are distinct fundamental descriptions of reality differing only by

---

[65] Or perhaps a stronger claim which adds that there are no fundamental properties or relations other than R. Perhaps Dasgupta would allow ordinary quantification over universals.

[66] For discussions of further related views, see Turner (2011) on "predicate functorese" and Russell (2018) on a form of generalism according to which nonqualitative facts are indeterminate in a certain sense.

a permutation of individuals. For in scenarios differing by such a permutation, exactly the same general sentences are true. Thus the view would seem to be supported by Dasgupta's anti-individualistic argument, anyway, as well as by the hole argument, and perhaps by the argument from quantum statistics.

Quantifier generalism induces the sort of metaphysical vertigo noted in section 2.5. For it violates the principles that played such a prominent role in Chapter 2: that existentials are grounded in their instances, that existential sentences are never fundamental (or at least, are fundamental only if their instances are as well), and so forth. (It also violates analogous principles about the grounding and fundamentality of universally quantified facts, such as that $\forall x F x$ is grounded in all of its instances taken together, perhaps together with a "totality fact".) Insofar as those principles are secure—the support for them in Chapter 2 was inconclusive—they provide a reason to prefer algebraic generalism over quantifier generalism. In any case, the arguments to be given in the following sections don't depend on the existentials principles, and thus threaten both views.

3.14.1 *Holism and expressive resources*

Dasgupta's view is holistic in a certain sense.[67] Suppose we want to describe a complex system, requiring two or more attributions of relations. Such attributions can be "linked" given individuals, by the recurrence of names, but as we saw in section 3.9, they cannot be linked in the multilocational bundle theory. Now, as we saw, Dasgupta can describe such complex systems. But he cannot do so with a series of distinct sentences—that too would omit the linkage. Dasgupta can describe complex systems only by constructing a single sentence that describes the entire system.

To illustrate, consider what the believer in individuals would describe with the pair of attributions Rab and Sab. In place of the first attribution Dasgupta can offer:

ccR obtains $\qquad$ (predicate logic equivalent: $\exists x \exists y R x y$).

In place of the second he can offer:

ccS obtains $\qquad$ (predicate logic equivalent: $\exists x \exists y S x y$).

But in place of the pair of the attributions Rab and Sab, Dasgupta cannot offer the pair of his replacements. For that pair would leave out the fact that, as the individualist would put it, the two situations involve the same individuals, and the same direction of holding of the relations ($\exists x \exists y R x y$ and $\exists x \exists y S x y$ don't imply $\exists x \exists y (R x y \wedge S x y)$). Instead Dasgupta must supply a single sentence that describes the entire situation: the term functorese translation of $\exists x \exists y (R x y \wedge S x y)$. And if he wants to go on to describe some larger system of which this situation is a part, he cannot simply add some further sentences describing the remaining parts of

---

[67] For discussion of issues in this vicinity see Hawthorne and Sider (2002, pp. 62–3), Dasgupta (2009, pp. 55–6), and Turner (2011, section 4.2.3).

the larger system (and their relations to the smaller situation). Rather, he must begin anew, and provide a single sentence for the larger system.

All this is true of quantifier generalism as well. A complex system must be described using a single quantified sentence, since a collection of quantified statements describing the proper parts of the system will omit any linkages between those parts.

Both forms of generalism, then, are "holistic", in that the whole truth about a system is a single fact, and does not reduce to a collection of multiple "smaller" facts.

This holism leads to my main objection. Since both quantifier and algebraic generalism require the whole fundamental truth about the largest system, the system consisting of the entire universe, to be describable by a single sentence in a fundamental language, this requires the fundamental language to have strong expressive resources—infinitary quantification and conjunction, say—if the universe is infinite.

The individualist, on the other hand, has no such commitment to strong expressive resources. For the individualist, the totality of facts stateable in a first-order language might well completely characterize the world. For each individual, a, the individualist can admit simple fundamental facts concerning a (Fa, Rab, ...). Given infinitely many entities there will be infinitely many such facts; but these facts will "link up" with one another via recurrence of the individuals in them.

Now, there are various well-known arguments that we need logical resources beyond those of first-order logic. For example, George Boolos (1984) argues that we need plural quantification to give an attractive set theory and to do natural language semantics. But the generalist's need for these further resources is quite different: they are needed at the fundamental level, and they are needed simply to state the physical facts about an infinite world. One might not accept arguments like Boolos's. Or one might regard such arguments as failing to establish the need for such resources at the fundamental level, as I argue in Sider (2011, section 9.15). Or one might think that such arguments demonstrate the need for further fundamental logical resources in logic, but not in physics. (I myself would reject this bifurcation, though.) Or, the further logical resources needed by the generalist might be more powerful than those established by the arguments. Given any of these outlooks, the added logical resources demanded by generalism go beyond what is demanded elsewhere. And so, I say, we have a reason from parsimony against each of the two forms of generalism. Each requires powerful logical concepts to articulate the fundamental facts—concepts that are not required by individualism.

To be sure, like all arguments from parsimony, this is merely prima facie, and could potentially be outweighed by the arguments in favour of generalism. I questioned the arguments for structural realism in section 3.4, and will reply to

Dasgupta's argument for generalism in 3.15 below; but the case against generalism is, I think, less decisive than the case against monism, say (and certainly less decisive than the case against the bundle theory). Generalism and individualism are both live possibilities for being the fundamental theory of the world, and the considerations that bear on the choice between them are tentative, contentious, and intertwined with many difficult issues, as we will see.

Let's make the argument for the need for strong expressive resources in more detail, in various cases. First, for purposes of illustration, pretend that reality consists solely of the positive integers, related by successor, addition, and multiplication.[68] The truths available to the generalist are the name-free sentences in the language of arithmetic, or their translations into term functorese. Now, the totality of the former sentences is not "categorical": the set of these sentences has nonstandard models that are not isomorphic to the standard model. So consider a "possible world" in which the "numbers" are structured as in one of the nonstandard models. Exactly the same name-free first-order sentences are true in this world and in a world in which the numbers are "normal", and thus the worlds share the same fundamental facts given either form of generalism, provided neither introduces extra logical resources into the language used to state the fundamental facts. But, premise:[69]

Supervenience The totality of facts supervenes on the totality of fundamental facts.

Thus it cannot be that the fundamental facts are all and only those recognized by the quantifier generalist. (The believer in individuals, on the other hand, can distinguish the worlds: they have nonisomorphic collections of fundamental facts about particular entities.) It may be objected that mathematical facts are necessary and so supervene on anything. But this is a red herring; we could rerun the argument with an infinite collection of physical objects structured by physical relations playing the role of the successor, addition, and multiplication relations.

A quantifier generalist might respond by introducing more powerful logical ideology—monadic second-order quantifiers, say—into the language used to state the fundamental facts; and perhaps Dasgupta could say something similar. Monadic second-order arithmetic is categorical: its models are all isomorphic to the standard model, and thus completely equivalent by the quantifier generalist's lights (since a permutation of entities is a non-difference according to her). The embrace of second-order quantifiers, or some corresponding term-functor-theoretic addition, might be defended as a small but tolerable affront to parsimony.

---

[68] In this case we won't need to appeal to the assumption that the quantifier generalist needs a single sentence to capture all of reality.

[69] This principle is inadequate as a full statement of the sense in which the fundamental facts are "complete" (section 2.5.2), but its use in the present context doesn't require it to play that role; it must merely be true.

But generalists won't be able to stop with just these additions. The truth about arithmetic is comparatively easy to express in a name-free language, because the natural numbers are so simple. But consider the whole range of possibilities for what happens in space. For simplicity, imagine that the metaphysician's favourite physics, a Democritean physics in which each point may be "on" or "off", is true. Since there are continuum-many points, each of which may be on or off, there are more than continuum many possible worlds of this sort.[70] By the Supervenience principle, no two of these worlds can share exactly the same fundamental facts; thus, there must be more than continuum-many sets of potential fundamental facts. But the quantifier generalist's language cannot express that many potential fundamental facts, if no infinitary resources are added. For its sentences would then always be finite in length; its vocabulary is countable (assuming that the set of simple universals is at most countably infinite); the set of sentences in any such language is countably infinite; there are only continuum-many sets of sentences in such a language. In order to describe such worlds (and also, presumably, our own world), a quantifier generalist seems to need an infinitary language allowing infinitary quantification and conjunction, in order to state a ramsey sentence with continuum-many existential quantifiers and (at least) continuum-many conjuncts.[71] And Dasgupta would need some sort of term functorese substitute for such a sentence.

Why doesn't this argument apply to the individualist? Because the individualist's language for expressing fundamental facts does not have a countable vocabulary; it can contain a name of each entity. Or, to put it nonlinguistically, in terms of facts, there are more than continuum many components of the fundamental facts in such a world, if particular entities can be components of fundamental facts as the individualist thinks; whereas if either form of generalism is true, only a countable set of constituents are available to occur in fundamental facts.

Thus there is pressure on quantifier generalists to enhance their logical ideology, to include, for instance, infinitary quantification and infinitary conjunction; and there is pressure on algebraic generalists to include corresponding enhancements. And as Jason Turner pointed out, there is also a further question for

---

[70] Hitch: translations and rotations of the pattern don't change the world, by the generalist's lights. Solution: just consider the points in some finite region, and add "reference objects" elsewhere in the world.

[71] Might set theory demand even greater expressive resources? Suppose the only way for the quantifier generalist to describe all the truths about the universe of sets was by "brute force"—by its Ramsey sentence $\exists x \exists y \ldots (\sim \exists z\, z \in x \wedge \ldots)$. Then *really* strong expressive resources would be needed: the sentence couldn't be formulated in any of the standard infinitary languages, not even in the "language" $L_{\infty,\infty}$. But McGee (1997) has proved that the axioms of second-order ZFCU plus an axiom saying that the urelements form a set is in a sense categorical, in that roughly any two models of this theory that share the same domain are isomorphic. So perhaps the quantifier generalist could get by with a finite (albeit second-order) language here. (Something like this problem would arise again if very strong principles of "recombination" were true that implied the existence of possible worlds in which the domain of concrete objects was proper-class size.)

algebraic generalism of whether such enhancements exist, of whether any attractive infinitary version of term functorese exists. What would the infinitary analogues of the padding and inversion functors be?[72]

It might be objected that the individualist is in no better shape than the generalist, since positing infinitely many individuals in ontology—or including infinitely many names in a fundamental language—adds complexity as well. The anti-generalist argument assumes that infinitary logical concepts offend against Ockham more than infinitely many individuals; but is this assumption justified?

It's often assumed that the mere *number* of individuals is less important to parsimony than the number of *kinds* of individuals one posits.[73] If this is right, and if, further, the infinitary logical concepts should be classified with "kinds" for these purposes, then the argument stands. But is it right; and if it is, why is it right? I think it is right, and that the reason—or part of the reason, anyway—has to do with laws: what is most important in parsimony is keeping the laws simple.

Laws don't mention particular individuals by name. Newton's laws of motion don't mention me, the Earth, Edna the electron, or any particular entity; they say, rather, that *any* (physical) object will behave in certain ways. But laws *do* mention particular *kinds* by name. The laws of classical physics, for example, mention both charge and mass by name (rather than quantifying generally over

---

[72] Turner made this and other helpful points in an APA commentary, and has pursued the matter further in Turner (2021). Dewar (2019a) introduces a form of structuralism that is like Dasgupta's but employs a different variable-free approach to predicate logic, one based on cylindrical algebras. Unlike Dasgupta's approach, Dewar's approach presumably generalizes smoothly to the infinitary case. For the operation c_i in cylindrical algebras may be indexed to arbitrary ordinals (rather than natural numbers, as in the main case that Dewar discusses); and the relations in those algebras may be thought of (intuitively) as relations of sequences indexed by arbitrarily large ordinals (rather than ω). However, note two features of Dewar's approach. First, the fundamental physical facts involve indexing to ordinals. Second, for each of Dasgupta's fundamental finite-placed relations, Dewar's algebra contains an infinite class of countably infinitely placed relations which differ from one another only over, intuitively, which argument places the fundamental relation is applied to. To a fundamental binary relation R there corresponds a relation that (intuitively) applies to an infinite sequence $x_1, x_2, x_3, \ldots$ if and only if $R(x_1, x_2)$, a relation that applies to the sequence if and only if $R(x_1, x_3)$, a relation that applies to the sequence if and only if $R(x_2, x_3)$, and so on. These features of Dewar's account are what enable the generalization to the infinitary case; but such artificial and redundant structure seems out of place in an account of fundamental reality. Dasgupta avoids this artificiality and redundancy since his analogue of c_i is his term functor c for existential quantification, which is not indexed; and the work of "lining up argument places" that is accomplished in Dewar by redundant infinitely placed relations is accomplished in Dasgupta by the padding and inversion functors. Thus the very features that keep Dasgupta's account from generalizing to the infinitary case are those that avoid the artificiality and redundancy. Note, however, that Dewar himself would not find the artificiality and redundancy objectionable, since he elsewhere (2015; 2019b) defends an across-the-board rejection of demands to eliminate such artificiality and redundancy. In the terminology of Chapter 5, he embraces quotienting. Thus the present issue—like a great many others—depends crucially on whether quotienting is acceptable.

[73] 'The default reading of Occam's Razor in the bulk of the philosophical literature is as a principle of qualitative parsimony' (Baker, 2016); see also Lewis (1973, p. 87); though see Nolan (1997) for dissent.

all properties) and say distinctive things about them. This means that when we posit new individuals, we don't in general posit new laws, whereas when we posit new kinds, we do generally posit new laws governing those kinds. So if keeping the laws simple is what is most important in parsimony, then reducing the number of kinds is more important than reducing the number of individuals.[74]

Now, logical laws mention particular logical constants (such as conjunction and existential quantification) by name. Dasgupta's additions are thus in this way like "kinds": they will enter into distinctive logical laws, just as the usual logical constants have their own distinctive laws, and just as properties of mass and charge and the rest have their distinctive laws of nature. So they should be treated like physical kinds, and unlike particular individuals, when it comes to parsimony.

It might be objected that infinitary expressive resources are needed at the fundamental level anyway, even if we reject all forms of generalism. One argument for this conclusion would be this: (i) a universally quantified fact $\forall x F x$ is grounded in the plurality of its instances plus a "totality fact"; (ii) the totality fact is $\forall x (x = a \lor x = b \lor \ldots)$, where all entities are listed; and (iii) the totality fact is fundamental. Then infinitary disjunction would be needed to state some of the fundamental facts, whenever there are infinitely many entities.

But the suppositions needed to state this problem are not stable. Given (ii), the totality fact is universally quantified, and thus given (i) it partially grounds itself, which is generally assumed to be impossible.

Fine (2012, p. 62) argues (for related reasons) that totality facts are not universally quantified, but rather have some other nature. If he is right, totality facts themselves might be argued to involve infinitary resources to which even nongeneralists are committed. Whether totality facts would require infinitary resources depends on subtle issues about totality facts. For instance, if they consist of the instantiation of an infinite-placed relation of totality by all the individuals $a, b, \ldots$ then this infinite-placed relation is required; but if they consist instead of the instantiation of a monadic property of totality by the mereological sum of all individuals (compare Armstrong (2004, section 6.2)) then no infinitary resources are required. But in any case, the argument is now relying on details of the ground-theoretic approach to quantified sentences which needn't be accepted. A fan of "grounding-qua" (section 2.5.2) might say that universally quantified statements are grounded in their instances qua their being all the instances, thus dispensing with totality facts. And I myself reject the standard account of the grounds of logically complex truths altogether (section 2.5.3). Universally quantified facts needn't be grounded in totality facts since universally quantified facts can themselves be fundamental facts.[75]

---

[74] Thanks to Tom Donaldson here.

[75] Universally quantified facts with nonfundamental constituents (say, nonfundamental properties) do need grounding, but they can be grounded in fundamental universally quantified facts that lack such constituents.

The objector might regroup. Rather than trying to establish an independent committment to infinitary resources via consideration of totality facts, she might instead try to establish such a commitment more directly: (i) it is a fact that $\forall x(x = a \vee x = b \vee \ldots)$; (ii) if this is a fundamental fact, then infinitary disjunction is needed to state the fundamental facts (assuming there are infinitely many objects); (iii) but if it is not a fundamental fact then it must be grounded, or otherwise rest on, fundamental facts; and it is hard to see how that could be, if no "infinitary fundamental facts" are admitted.

Now, whether (iii) is correct depends on some difficult questions about the sense in which the fundamental is "complete" (recall section 2.5.2). A defender of grounding-qua, for instance, might claim that $\forall x(x = a \vee x = b \vee \ldots)$ is grounded in the facts $a = a$, $b = b$, $\ldots$, qua those being all the identity facts there are. But the real problem is the argument's assumption in step (i) that there is an infinitary fact about all objects, $\forall x(x = a \vee x = b \vee \ldots)$. Making that assumption in the present context is like trying to convince a modal sceptic with an argument that assumes that there are modal facts. Unless some argument is given that such a fact exists, step (i) may simply be denied.

One final point about all these attempts to establish an independent commitment to infinitary resources: even if they succeed, they only establish a commitment to certain limited infinitary resources: infinite disjunction in some of the attempts, an infinitely placed predicate in one attempt. But the generalist needs a further infinitary resource: infinitary quantification (or some algebraic analogue).

In a critique of my book *Writing the Book of the World*, Fine distinguished the "D project", that of completely describing the world, from the "E project", that of *expressing* facts in the most fundamental terms. He says:

> We can easily bring out the difference between the two projects with the case of disjunction. I can say 'p or q' and it is not clear that this can be said except by using disjunction or the like. But suppose now that I correctly describe the world by means of the sentence 'p or q'. Then the use of 'or' is dispensable, since I can alternatively describe the world by means of p or q, depending upon which is true. Thus even though 'or', or the like, may be indispensable for saying what we can say, it would not appear to be indispensable for describing what we can describe.
>
> Fine (2013, p. 730)

It might seem at first that this distinction is relevant to my criticism of generalism. My criticism is that the generalist needs infinitary logical concepts, but Fine's distinction suggests the idea that logical concepts are never needed for the D project; and it is surely the more "worldly" D project that generalism is meant to address. But in fact I don't think that Fine's distinction would help here, since it is in the D project, not the E project, where the generalist encounters the need for infinitary quantification. The need for infinitary facts was not expressive, but rather was for giving a complete description of the physical world.

3.14.2 *Holism and scientific explanation*

A second possible objection to generalism (of either sort) concerns the impact of holism on explanation.

How to decide what is fundamental? In section 1.8 I said that the fundamental concepts we posit should enable simple and strong laws of nature. And the generalist can easily comply with this. For laws are purely general statements, and thus can be stated in terms that the generalist regards as fundamental.

But perhaps our epistemology for fundamentality should require something further. Enabling simple and strong laws might be seen as a special case of enabling an attractive reconstruction of scientific practice as a whole. According to a stronger requirement, the fundamental concepts we posit should not only be capable of stating simple and strong laws, but should also be capable of stating good scientific explanations.

Compliance with this stronger requirement is not so easy for a generalist, since explanations are not always perfectly general; often they concern particular matters of fact. Suppose there is a fundamental law that:

(L) Whenever a particle that is F bears R to another particle, that other particle is G,

and consider a very simple explanation, concerning a certain pair of particles a and b:[76]

(E) Particle b is G because particle a is F and bears relation R to particle b.

(E) mentions particular objects, and thus is not stated using fundamental concepts according to generalism. Is there any corresponding explanation in perfectly general terms?

The generalist can state a generalization about explanation that in some sense encompasses (E): for any particle x that is F and bears R to another particle y, y is G because x is F and bears R to y. But this generalization is distant from what was achieved with (E), since (E) was offered as an explanation of what happened in a particular situation, the situation involving a and b. For similar reasons, an existential generalization corresponding to (E), namely that there exist particles x and y such that y is G because x is F and bears R to y, would not achieve anything like what (E) achieves. For there may be many particles x and y with these features, not just a and b; the existential generalization is insufficiently specific.

The closest fundamental counterpart to (E) available to the generalist is a strengthened version of the existential generalization just mentioned:

(E_G) There exist particles x and y such that x has certain features, y has certain other features, and y is G because x is F and bears R to y,

where the "certain features" are had only by a and the "certain other features" are had only by b. These features cannot in general be intrinsic, since a and b

---

[76] Perhaps the explanans should explicitly mention (L). I do not mean to assume anything about the general form of explanations.

needn't be intrinsically unique. They must often be extrinsic, picking out a and b by their relationships to other objects, for example 'x is exactly such-and-such distance from some z which has feature H, and y is exactly so-and-so distance from some z' which has feature H''. If the world is sufficiently large and varied, these conditions might need to be complex, or involve spatial relations to distant objects (and if the world is symmetrical in certain ways then the conditions would not exist at all).

The concern, then, is that (E_G) won't be as attractive an explanatory claim as (E). From the fundamental point of view of the generalist, there cannot be the sorts of simple explanations of particular occurrences that we offer in ordinary scientific practice.

Is this a serious cost of generalism? I'm not sure. On one hand, one might insist that not only should laws look good, by ordinary scientific standards, when viewed through the lens of fundamentality, but also, explanations should look good, by ordinary scientific standards, when viewed through the lens of fundamentality. But on the other hand, in the counterpart explanation (E_G), the added complexity is only in the "certain features" used to single out a and b. After those features are mentioned, the final part of (E_G) is just like (E):

(E_G) There exist particles x and y such that x has certain features, y has certain other features, and **y is G because x is F and bears R to y**.

Moreover, there is a sense in which all the explanatory action occurs in this final part, since the formula in it does not depend on the "certain features"; it would be true of any pair of objects where the first is F and bears R to the second.

3.15 Against Dasgupta's argument

The objections to generalism have not been decisive. Thus it matters how powerful Dasgupta's argument for that view is. Let's look at it more carefully.

Dasgupta's argument was that the reasons to reject absolute velocities are also reasons to reject individuals. Those reasons were physical redundancy and empirical undetectability, and they raise different issues.

From a certain "realist" point of view, the complaint about empirical unde-tectability carries little weight.[77] For the realist, there is nothing whatsoever wrong with a theory that implies the existence of facts that we cannot—in various senses— know. We should embrace the external world, the existence of unobservables, and facts about the distant past, even if in various senses we cannot know about these things. Now, Dasgupta does not mean to be denying this broadly realist thought. His complaint is only about a specific and nuanced sort of empirical undetectability. Still, the force of this complaint is undermined by the realist thought. If nothing is wrong with unknowability per se, why should there be something wrong with one particular sort? From the realist point of view, the

[77] See also North (2018, section 2).

complaint about unknowability—both in the case of absolute velocities and in the case of individuals—is entirely unmotivated.

When a theory implies the existence of unknowable facts of various sorts, this is often a sign that something nonepistemic has gone wrong. The claim that there exist little green men who disappear whenever we attempt to observe them implies unknowable facts, but there are entirely nonepistemic reasons for rejecting it: it posits entities that play no role in explaining the evidence. Explanatorily idle elements and other such theoretical defects generally lead to unknowable facts; absence of such defects usually (though not always) leads to testable consequences; that is why unknowable facts are a sign that some theoretical defect might well be present. But unknowability does not *always* indicate a defect; theoretical improvement does not always go hand in hand with easier epistemology. Indeed, consider the paradigm case of an appeal to the theoretical virtues: Bertrand Russell's (1912, chapter 2) argument that we have reason to believe in the external world because its explanation of our sensory experience is simpler than the "idealist" hypothesis that nothing exists other than sense data. The external world hypothesis is usually thought to imply the existence of *more* unknowable facts than the idealistic hypothesis.

From this point of view, the empirical undetectability of absolute velocities in Newtonian gravitational theory is not, on its own, a strike against them; at best it is a sign that embracing them would result in some other theoretical defect—physical redundancy, perhaps. Thus in my view—though Dasgupta disagrees (2021; 2016b)—it is the complaint about physical redundancy, not the complaint about empirical undetectability, that is the stronger of the two proposed reasons to reject absolute velocities. Physically redundant posits in a theory are "wheels that turn" without being properly incorporated into the rest of the theory's explanatory mechanisms. Absolute velocities would not be *completely* inert in Newtonian gravitational theory, of course, since absolute velocities at one time would affect absolute velocities at other times. But, so the thought goes, since that's *all* they affect, they are insufficiently integrated with mass, relative velocity, etc. Simply dropping them from Newtonian gravitational theory sacrifices nothing of explanatory value, and thus results in a superior theory, even from the "realist" point of view. As it's often put, absolute velocity structure should be rejected because it is excess structure that is not needed for the dynamics of Newtonian gravitational theory to make sense (Earman (1989, p. 46); North (2009, section I)).

So our question is whether this complaint about physical redundancy can be made about individuals. In fact this is far from clear, at least if the issues are viewed through the lens of concept-fundamentality. For through that lens, it is far from clear that individuals are physically redundant in the relevant sense—an insufficiently integrated explanatory posit. Given how integrated the concept of an individual is within individualistic theories, one cannot simply "scoop out" the individuals and leave the rest of the theory intact. The individuals-theoretic fundamental concepts—names, quantifiers, predicates—are part of every statement the theory makes. The conceptual change Dasgupta wants us to make, namely replacing these concepts with names of universals and his term functors, is not a *deletion* of structure from the theory, but rather an *exchange* of one sort of structure for another. The complaint about "redundant" structure is ultimately an Ockhamist one: if we can simply *delete* a component of a fundamental theory without sacrificing anything of explanatory importance, we should do so, since the deleted component wasn't doing any distinctive work that needed to be done. But if one component can be *exchanged* for another without sacrificing anything of explanatory importance, this doesn't show that the exchanged component was redundant; it just shows that there's a different way to get the same job done. One might of course argue that the alternate way to get the job done is better, but the mere possibility of the exchange doesn't establish this.

The key point here is that the judgement that a theory has excess or redundant structure is, when viewed through the lens of concept-fundamentality, a judgement about a privileged statement of that theory, complete with distinguished putative fundamental concepts, distinguished formulations of laws in terms of those concepts, and so forth. If there is excess or redundant structure in such a theory, that must be a fact about those privileged concepts and how they are utilized in the laws. The mere fact that a certain "aspect" of the theory (such as that of which particular individuals play which roles at a certain time) is isolated from other aspects (in the sense that, for example, a permutation of individuals over roles at a time won't affect anything at later times other than which individuals play which roles then) does not show that the theory contains redundant structure, if the aspect in question does not correspond to a single distinguished element in the theory—a distinguished fundamental concept, or a distinguished law, say.

My "isolated aspects" are like *symmetries* (of the laws), and the point can be put in those terms as well: the mere existence of a symmetry does not show that the theory contains redundant structure in the sense of dispensable distinguished elements. Compare Arntzenius (2012, p. 178):

I do not deny that it can be good to have a theory which has fewer 'symmetries'—where by a 'symmetry' I mean a transformation which leaves the dynamics and the phenomena invariant. Getting rid of apparent redundancies in one's formalism is indeed, other things being equal, a good thing, for it reduces one's commitments—*but only if it leads to a simpler (empirically adequate) theory.*

(In Chapter 5 we will discuss "quotienting", which rejects concept-fundamentality and which does allow "scooping out" arbitrary aspects of a theory's structure. Thus we have another illustration of the importance of the choice of metaphysical tools to these issues.)

It must be conceded that this way of viewing physical redundancy creates a problem for all judgements of physical redundancy, including the judgement that absolute velocities are physically redundant in Newtonian gravitational theory. Suppose that in Newtonian gravitational theory we are choosing whether to adopt Galilean or Newtonian space-time, and in particular choosing which of the following sets of fundamental concepts to adopt:[78]

Galilean concepts spatial-distance-at-a-time, temporal distance, affine connection.

Newtonian concepts spatial-distance-at-a-time, temporal distance, affine connection, same-place-as.

This particular choice is straightforward: the Galilean concepts are a proper subset of the Newtonian concepts, and for that reason are preferable on grounds of parsimony. Moreover, let us suppose, the laws of Newtonian gravitational theory can be formulated simply (and attractively) using only the Galilean concepts—same-place-as doesn't even appear in those laws. Given this, the same-place-as relation is redundant in a very clear sense, and may simply be dropped.

However, the geometry of Newtonian space-time can be characterized using different fundamental concepts from those just mentioned. Consider these concepts for that geometry, mentioned in a co-written dialogue by Dasgupta and Jason Turner (2016):

Alternate Newtonian concepts cross-time-spatial-distance, temporal distance.

The choice between these and the Galilean concepts is far less clear. Neither set of concepts is a proper subset of the other. The laws of Newtonian gravitational theory, when formulated using the Alternate Newtonian concepts, would now employ both of its concepts; thus that set of concepts would not include an element that simply doesn't occur in the laws.

Thus a single theory can be given multiple formulations, corresponding to multiple choices of fundamental concepts, which lead to different judgements about redundancy. One obstacle this poses to judgements of redundancy viewed through the lens of concept-fundamentality is that of which formulation to consult when making the judgement. Now, this isn't really an obstacle, since from the point of view of concept fundamentality, the "formulations" are just different theories. (This of course creates an epistemic problem, which we'll discuss in Chapter 5.) But a second obstacle arises: for some of these formulations/theories, it's unclear what the judgement should be. Are the Alternate Newtonian concepts superior or inferior to the Galilean concepts?

Dasgupta and Turner bring up the alternate Newtonian concepts in an attack on the idea that absolute velocities are to be rejected because they give rise to ideological complexity. As they point out, the judgement of complexity isn't clear since Newtonian space-time can be given a formulation—based on the Alternate concepts—which has fewer fundamental concepts (two) than a formulation in terms of the three Galilean concepts. Now, as Goodman and others have pointed out, one can always artificially reduce the number of primitive expressions in a

[78] I'm simplifying by ignoring details of how to incorporate the quantitative aspects of these notions.

theory by cooking up appropriate definitions.[79] So evaluating ideological simplicity by simply counting fundamental concepts was never an option. But the point remains that it is far from clear how to make judgements of ideological simplicity. And for similar reasons, it's far from clear how to make judgements of redundancy, viewed through the lens of concept-fundamentality.

The fan of concept-fundamentality has no choice but to concede that these matters are complex and multifaceted and (at present anyway) nonalgorithmic. It would be foolish to jettison the entire approach simply because of the messiness. What we are talking about here, after all, is the messy epistemology of theory choice, and there's no reason to suppose that the correct superempirical principles governing such choices are tidy (or even that there's always a fact of the matter about them). The realist is just stuck in this muck, and had better learn to live with it.

As for life in this muck, one might invoke simplicity comparisons between individual notions (as opposed to between entire ideologies). One might argue, for instance, that the notion of cross-time-spatial-distance is more complex than spatial-distance-at-a-time. It's certainly complex when viewed from the perspective of what I originally called the Newtonian concepts, since one can reductively define the distance between nonsimultaneous p_1 and p_2 in Alternate Newtonian terms, as the distance between p_1 and the point simultaneous to p_1 that is at the same place as p_2. But what I am suggesting is that cross-time spatial distance might be more complex in some intrinsic or absolute sense (and not just relative to the Newtonian concepts), that facts stated using that notion are in a sense richer, contributing to a more richly structured world, than facts stated using the notion of spatial distance at a time.

It might also be useful in the muck to simultaneously evaluate for simplicity a proposed set of concepts and set of laws (and by the latter I mean distinguished formulations of laws in terms of the proposed concepts). For instance, there might be a sense in which an attractive formulation of the laws of Newtonian gravitational theory only "looks directly at" spatial distances at a time, and never at cross-time spatial distances, which could be taken as a strike against the latter. (This seems clear in the case of the law of gravitation, but I'm not sure how to think about what the law of motion "looks at".)

3.16 How far to go?

How far would the demand to remove undetectable or redundant structure lead, if given completely free rein?

The demand led Dasgupta to eliminate individuals. But a parallel demand would seem to lead to the elimination of scientific properties as well. Permuting charge and mass within a combined Newtonian gravitational theory + Classical electromagnetism theory, say, might be argued to have no observational effect, as

---

[79] See Goodman (1951, chapter 3) for a discussion of some of these issues.

Turner (2016a) points out. Likewise, one might argue that individual properties of charge and mass are physically redundant in that theory, because an initial permutation of charge and mass would result in correspondingly permuted outcomes. Thus arguments parallel to Dasgupta's would seem to imply that we ought to eliminate properties like mass and charge, in addition to individuals.

Would it even make sense to eliminate that much? It clearly would if the "elimination" were modelled on quantifier generalism. On this view we don't really eliminate either individuals or (scientific) properties in the sense of denying that they exist; rather, we say that, in a fundamental description of reality, we cannot name *particular* individuals or properties, nor can we use particular physical predicates. (Instead of the parenthetical 'scientific' we could instead write 'sparse'; the contrast is to "abundant" properties in Lewis's (1986a, p. 55 ff) sense.) We must rather speak of both individuals and properties generally. We cannot say that a has a certain charge and b has a certain mass, but we can say that there exist individuals x and y and scientific properties p and q such that x has p and y has q. (If the resulting view feels too bare, a little more meat could be added to the bones: ideology from a robust theory of lawhood, such as Armstrong's 'neccesitates' (1983), or a one-place sentence operator 'it is a law that', could be spared Ramsey's axe, not quantified out. Thus fundamental facts could look like this: there exist properties p and q and an individual x such that p necessitates q and x has p.) This view makes sense—at least, insofar as quantifier generalism (with its denial that existentials are grounded in their instances) makes sense.

Rejecting both individuals and properties in Dasgupta's way seems initially *not* to make sense. Dasgupta's method for eliminating individuals is to replace any theory about individuals with the statement that a certain complex zero-place universal obtains, where that universal is constructed (via the term-functors) from the basic universals of the theory. But now we seem to be trying to eliminate those basic universals as well. How then will the complex universal be constructed?

To make the problem clearer, as well as to point the way towards a solution, let us examine Dasgupta's conversion of sentences about individuals into term functorese a little more carefully. Begin by noting that the *starting* point of the conversion cannot contain any names of individuals; it may only contain quantifiers over individuals. Thus it might look like this:

(Start) $\exists x F x$.

The next step is to convert the predicates to names of universals: the predicate 'F' is converted into a name 'F-ness' of a universal. (Here it is important to be more careful about grammar, so I use a distinct expression for the universal.) Now, suppose we want to write down an *intermediate* sentence at this point, after the conversion of predicates to names but before the final translation into term functorese. We can't do this by flat-footedly substituting the names of universals for the corresponding predicates, since the result would be ungrammatical: (Start) would become '$\exists x F\text{-ness}\,x$', which isn't grammatical since 'F-ness x' is in sentence position but is a string of two singular terms. But what we can do is introduce a predicate 'has' (or 'instantiates'), and speak of individuals having properties. Thus we have our intermediate sentence:

(Intermediate) $\exists x$ has$(x, F$-ness$)$.

('something has F-ness'). This then can be converted to term functorese, arriving at the *final* sentence:

(Final) cF obtains.

The problem we are confronting can now be seen. Just as the starting point for the conversion above had to be a sentence lacking names of particular individuals, our starting point for the extended Dasguptan elimination must now, in addition to lacking names of particular individuals, also lack names of, or predicates for, particular properties, since these are also to be eliminated. We may quantify over the properties to be eliminated (and over individuals), but we cannot name them. A very simple sentence of this sort might look like this:

(Start$_E$) $\exists x \exists p$ has(x, p).

(The subscript 'E' is for 'extended', the extended Dasguptan elimination.) The problem now is this: we need an intermediate sentence containing names of universals, which can subsequently be converted to a term functorese statement about a complex built out of those universals. But what will the intermediate sentence about universals be, when the starting sentence (Start$_E$) has been stripped of all names of particular properties?

When the problem is put this way, a solution comes into view. In the original example, the starting sentence (Start) didn't contain any names of particular properties either. But it did have a predicate, 'F', which in the move to (Intermediate) got converted to the universal 'F-ness'. In the present case, (Start$_E$) also contains a predicate, 'has'. Thus a solution would be to reify having, the relational tie between universals and particulars, just as earlier we reified F-ness. That is, the intermediate sentence for the extended Dasguptan elimination would be about a universal H of having.

Now, in order to construct this intermediate sentence we can't simply replace 'has' in (Start$_E$) with H, since the result would be ungrammatical, as it was above. What we must do is introduce a further "higher-level" *predicate* for having, in addition to the name of the lower-level universal H of having. To avoid confusion, let this new predicate be 'bears'. The intermediate sentence is then:

(Intermediate$_E$) $\exists x \exists p$ bears(x, H, p).

There is a point to this Bradleyan ascension: now we have a sentence containing a name of a universal, 'H'. Thus we have a suitable sentence for translation into term functorese; (2) goes into:

(Final$_E$) ccH obtains.

In this way, instead of making statements about complex universals built up from simple universals of mass, charge, and so forth (via the term functors), Dasgupta could make statements about complex universals built up solely from the universal H of having.[80]

(As before, if the view feels too bare, one could exempt from ramsification some concept from a robust theory of laws, only now the concept must be reified. We might accept, for instance, a universal N of Armstrongian necessitation—as Armstrong himself did. Example: starting sentence '$\exists p \exists q \exists x(\text{necessitates}(p,q) \land \text{has}(x,p))$' yields intermediate sentence '$\exists p \exists q \exists x(\text{bears}(p,N,q) \land \text{bears}(x,H,p))$', which yields functorese sentence '$ccc(\sigma pN \,\&\, \iota p \sigma H)$ obtains'.[81])

So: it does make sense after all to eliminate scientific properties in addition to individuals in Dasgupta's way. The crusade against undetectable and redundant structure *could* be carried this far. (Dasgupta expresses openness to a combined structuralism about individuals and properties in Dasgupta and Turner (2016).) But should it? I wonder where this will end. *Any* theory will contain "constants", expressions that are not ramsified away, which could then be permuted. There is a scenario in which conjunction and disjunction have exchanged logical roles; does this mean that a theory with logical constants for conjunction and disjunction has objectionably undetectable or redundant structure?

It may be responded—perhaps in the first place, regarding charge and mass, or later, in the case of conjunction and disjunction, say—that the imagined permutations are not possible, or, alternatively, that they are not really observationally distinguishable (e.g.: 'a world in which mass rather than charge behaves in such-and-such ways just *looks* different'). But it's hard to see why these responses would be any more successful here than they would have been at the outset, in the case of individuals.

It may be responded that the case of conjunction and disjunction, anyway, is different because we have no entities to permute. 'And' and 'or' are sentence operators, not names, whose semantic function is not to stand for truth functions or any other entities. But it's hard to see why this fact makes a difference. The permuted scenario can be described without reifying conjunction and disjunction: it is a scenario in which snow is white and grass is purple, in which grass is not purple, in which the proposition that A implies the proposition that A and B, and so on.

Jeff Russell made a nice distinction concerning the example of permuting conjunction and disjunction. There's a difference between saying that permutations of situations that are allowed by a theory are allowed *by that very theory, as originally stated*, and saying that permutations of allowed situations are allowed *by a permuted*

---

[80] H could be multigrade, to allow for the having of universals of variable number of places, and similarly for the predicate 'bears'; or there could be a family of having relations $H^2, H^3 \ldots$ of different 'adicies, and a family of predicates 'bears³', 'bears⁴',...

[81] ι is a further term functor which inverts the first two argument places of a relation. See also note 63.

version of the theory. Only the latter is true in the conjunction/disjunction case. If a theory allows $A \vee B$ as a possibility, then a "permuted" theory that exchanges $\vee$ and $\wedge$ in the logical laws will allow $A \wedge B$, but the original theory may not allow $A \wedge B$. With that distinction in mind, we might say: what's objectionable is for a theory itself to allow permuted possibilities; it isn't objectionable (or anyway, not in the same way) if a theory is such that its permutations allow the permuted possibilities. But now return to the case of permuting individuals in Newtonian gravitational theory: the original theory, without re-interpretation, allows the permuted possibilities: if Newtonian gravitational theory allows $\phi(a, b)$ then it allows $\phi(b,a)$. The crucial difference is that in Newtonian gravitational theory, construed so as to include the underlying logic, the laws contain no names of particular individuals, but do contain '$\wedge$' and '$\vee$' (as "constants"). Thus, from this point of view, one could continue to make Dasgupta's objection to individuals without pursuing the argument so far as to apply to conjunction and disjunction. In fact, one should stop even before the rejection of charge and mass, since that argument is like the argument against conjunction and disjunction, and not like the argument against individuals. Like '$\wedge$' and '$\vee$', 'Charge' and 'mass' occur as constants in the laws,[82] so there is no guarantee that if one's original theory (Newtonian gravitational theory + Maxwell, say) allows $\phi(\text{charge}, \text{mass})$, it also allows $\phi(\text{mass}, \text{charge})$; at best, a modified theory (which permutes 'mass' and 'charge') allows the latter. So Russell's way of thinking counsels Dasgupta to get off the boat right at the outset.

Is Russell right? It would seem to depend on whether the complaint about individuals is to their undetectability or (alleged) redundancy. If the former, then Russell's point seems wrong. For the complaint should then be just as compelling, regardless of whether we need to permute the theory in addition to the descriptions it allows. The argument would be this: if you accept the existence of individuals, or charge and mass, or an ideology containing conjunction and disjunction, then you must admit that there are these possibilities that are observationally equivalent: theory + outcome, permuted-theory + permuted-outcome. But suppose instead that the complaint is less epistemic (as I argued it should be), and is instead that there's some sort of redundancy in theories that posit individuals. Then Russell's point seems right, since the alleged badness pertains to the theory—the unpermuted theory—rather than to the hypothesis of individuals (or charge/mass, or conjunction/disjunction) per se.

---

[82] One might wonder whether this is contentious; might a nomic essentialist think that the laws are somehow variable with respect to properties as well as individuals? But this idea evaporates on closer inspection. The idea would be that the laws look like this: 'any property that ϕs is such that . . .'. But how to fill in ϕ? Not like this: 'Any property that plays role $\mathcal{L}$ in the laws is such that . . .', since this very statement is supposed to be the statement of the laws!

3.17 Generalist nomic essentialism

Our critique of nomic essentialism in Chapter 2 had a lacuna. Sections 2.1 and 2.4 relied on principles connecting ground and fundamentality to existential quantification: that existentially quantified facts are grounded in their instances, that existential facts are never fundamental facts, or (more weakly) that if an existential fact is fundamental, so are its instances. But those principles were given only inconclusive support (section 2.5). Moreover, reasons were given in section 2.5.3 for rejecting the first two principles if our central metaphysical tool is concept-fundamentality.

In light of the previous few sections, however, a new argument can be given to fill the lacuna. The natural way to reject the principles is to accept some form of generalism in the case of properties, whether quantifier or algebraic; but as we will see, the argument of section 3.14.1 against generalism about individuals can be made against these positions as well.

According to the form of nomic essentialism discussed in sections 2.1 and 2.4, facts attributing particular (scientific) properties (and relations) are not fundamental; properties must be quantified over in the fundamental facts. There is no fundamental fact that a has P; rather, there is a fundamental fact that a possesses some property that plays a certain nomic role: $\exists p(\mathscr{L}(p) \wedge a$ has $p)$. This is a form of quantifier generalism for properties.

In the previous section we saw what quantifier generalism for *both* individuals and properties would look like. On that view, neither particular individuals nor particular properties can be mentioned when expressing a fundamental fact. Quantifier generalism for properties alone, however, bans only names for properties; names of particular individuals are allowed.

My main objection in section 3.14.1 to quantifier generalism about individuals was that infinitary logical notions will be needed to state the fundamental facts. The objection also applies to each form of quantifier generalism for properties. It obviously applies to quantifier generalism for both properties and individuals; but in fact, it even applies to a quantifier generalism that is restricted to properties. On that view, a pair of individuals a and b could not in general be fully described using a sentence solely about a and a separate sentence solely about b:

There exist properties $p_1, \ldots$ such that $A(a, p_1, \ldots)$.

There exist properties $p_1, \ldots$ such that $B(b, p_1, \ldots)$.

That would leave out information about the relationships between the scientific properties possessed by a and those possessed by b. We would instead need to use a single sentence:

There exist properties $p_1, \ldots$ such that $A(a, p_1, \ldots)$ and $B(b, p_1, \ldots)$.

The same goes for any larger collection of individuals; and so for infinite collections of individuals we would in general need infinitary conjunction—though not infinitary quantification, unless there are infinitely many scientific properties.

(Infinitely many scientific properties would be required given a mixed absolutist approach to quantities (section 4.7.3), but presumably not given a comparativist approach (section 4.3).) The need for these infinitary resources persists even given individuals at the fundamental level because the linkage between distinct facts that would be supplied by recurrence of particular properties is missing.

For those nomic essentialists unwilling to reject the principles connecting existential quantification to ground and fundamentality, section 3.16 made available a further option: a combined algebraic generalism about both individuals and scientific properties using Dasgupta's term functorese. This view continues to require infinitary resources since it encompasses generalism about individuals. But since some nomic essentialists may not wish to eliminate individuals, we should ask whether the Dasguptan elimination of scientific properties could be combined with the acceptance of individuals.

In fact it can; here is one way to do it. As in section 3.16 the ontology includes no scientific properties, but it does include a having relation H and perhaps some nomic relation such as Armstrong's N. Moreover, we continue to translate any starting sentence into a statement of term functorese to the effect that a certain complex zero-place universal obtains. But our ontology now also includes individuals, and as a result we will also need new term functors to incorporate individuals into complex universals. One of these is a term functor π for "plugging", which takes a term for an $n + 1$-place universal U and a name for an individual, α, to form a term $\pi(U, \alpha)$ for the n-place universal that is the result of plugging up U's leftmost place with the individual named by α. The idea of the approach may now be illustrated with an example, using the terminology of section 3.16. Suppose we want to say that individuals a and b share some scientific property. We begin with the starting sentence '$\exists p(\text{has}(a, p) \wedge \text{has}(b, p))$' (the name of the shared property is quantified out, but the names of the individuals a and b remain). Next form the intermediate sentence by reifying having: '$\exists p(\text{bears}(a, H, p) \wedge \text{bears}(b, H, p))$'. And finally, translate to term functorese using, in particular, the plugging term-functor π: '$c(\pi(H, a) \& \pi(H, b))$ obtains'.

More would need to be said to develop this into a full theory.[83] But even from this example, it is clear that the concern about Dasgupta's approach raised in section 3.14.1, and again earlier in this section about the quantifier-generalist approach to nomic essentialism, applies here as well. Even though individuals are

---

[83] Dasgupta's functor c is used to pick out universals cU that allegedly do not "involve" or "presuppose" individuals. (I am granting him this for the sake of argument, though it might well be denied.) The defender of the present view should say the same about the use of c to eliminate quantification over universals, since those universals are to be eliminated. For instance, $c\pi(H, a)$ ('*a's having some property*') should be understood as not presupposing or involving universals (other than having). But individuals are *not* to be eliminated, and so a distinct term functor should be used to construct existentially quantified universals in which the quantification is over individuals. I will give this second term functor the suggestive name $\exists$; universals $\exists U$ *do* presuppose/involve individuals. Thus the universal $c\exists H$ ('*some individual's having some property*') presupposes the existence of an individual but not a property.

accepted, the truth about a pair of individuals cannot be given with two separate sentences. Rather, a single sentence must describe the pair; and similarly for larger, even infinite collections of individuals, which calls for some term functorese analogue of infinitary predicate logic.

3.18 Antistructuralist conclusions

A structuralist metaphysics of individuals has been proposed in a number of contexts: by ante rem structuralists in the philosophy of mathematics, by structural realists and their allies in the philosophy of physics, and by various people within general metaphysics. Although some criticism was made of arguments in favour of these ideas, our main focus has been on what a structuralist metaphysics of individuals might amount to.

As with many forms of structuralism, a modal understanding of structuralism about individuals is straightforward: the identities of individuals cannot (in one sense or another) vary independently from their qualitative structure—from the network of properties and relations they instantiate. But from a postmodal point of view, any such modal thesis needs some nonmodal underpinning. Our question has been: what might that underpinning be? What is the distinctively structuralist picture of the actual nature of individuals?

That picture has been elusive. It must in some sense downgrade the particular individuals in qualitative structures, relative to the structures themselves; but as we have seen, particular individuals play an essential role in characterizing those structures. A clear, disciplined account (as opposed to opaque slogans) is needed, which provides an adequate foundation for scientific and ordinary discourse, is theoretically attractive, and genuinely counts as structuralism.

We considered four main approaches: bare particulars, the bundle theory, monism, and generalism. (Indeterminate identity, weak discernibility, and "individuation" by relations were also discussed.) In each case we asked whether the view was viable, and whether it really counted as structuralism.

Bare particulars (a.k.a. moderate structural realism, in one sense) are individuals that stand in relations but have no fundamental properties. This is perfectly coherent metaphysics; no objection was given to its truth. But it is barely structuralism at all, for it fails to deliver the main prize: permuting the identities of individuals while holding qualitative structure constant remains a genuine difference at the fundamental level. ("Individuation" was critically discussed at this point.)

The bundle theory, in contrast, clearly counts as structuralism: it eliminates particular individuals from fundamental reality by identifying individuals with bundles of universals, thus delivering the prize. But its flaw is fatal: it cannot account for relations. Individuals play a crucial role in accounting for qualitative structures, that of "linking" multiple facts together to form that structure, and the bundle theory has no way to accomplish this linkage.

Monism downgrades the individuals in structures by claiming that all the facts about them are grounded in the facts about the structure itself, the biggest structure of all, the cosmos. This seems consonant with the structuralist picture, but a weighty objection was given. The facts about the cosmos that ground all other facts must be specifiable without reference to particulars, if the view is to be genuinely structuralist. But without an account of what these irreducibly cosmos-level facts are, and without an account of how they give rise to sub-cosmic facts, we do not yet have a determinate realization of the structuralist vision. Moreover, there are reasons to doubt that an attractive account of this sort can be given, one that enables simple laws of nature, for instance.

The fourth approach was generalism, which was discussed at great length. In one form, it says that the fundamental facts are existentially quantified; in another, due to Dasgupta, that the fundamental facts involve only complex properties and relations, whose structure is given algebraically. This view can be given a rigorous statement, it allows for simple laws of nature, and it captures as much of the structuralist picture as one could reasonably hope for. My objection was that it implies a holism that demands fundamental infinitary concepts to state the fundamental facts. I myself regard the objection as formidable (and as not being outweighed by convincing arguments in favour of generalism); but the objection involves some contentious assumptions and certainly is not decisive. Generalism is the most promising approach to structuralism that I know of.

But my chief message to structuralists in the philosophy of mathematics and philosophy of physics is to face up to the challenge of making metaphysical sense of their position. 'Relations without relata', 'Individuals are just positions in structures', 'Individuals are individuated by relations'—slogans like these are metaphysical ideas, but they are exceedingly cryptic ones. Faith that some sense or other can be made out of them is unjustified; there is no guarantee that any coherent position in the vicinity is both attractive and genuinely structuralist. The status quo, namely the standard approach to the foundations of physics and mathematics based on a face-value conception of individuals, is not mere prejudice, projection, or whim, but rather is a local equilibrium in conceptual space—perhaps the only attractive one.

4

Quantities

Quantitative properties and relations are those that come in degrees, such as mass, charge, and distance.[1] We talk about them using numbers, both to ascribe them to particular things ('object o is 5 kg') and also to state laws ('$F = ma$').

I will view the metaphysics of quantity through the lens of fundamentality. For, given the centrality of quantitative properties in physics, the question is about the ultimate nature of reality; and as we have seen in sections 2.3 and 3.11, in such contexts the appropriate postmodal tool is fundamentality, rather than ground or essence. This fundamentality-centric approach will lead to a focus on laws of nature. For competing accounts of quantity have distinctive implications about the laws of nature, via their distinctive accounts of the fundamental physical properties and relations that enter into those laws.

This chapter partly fits the larger pattern of studying the impact of postmodal metaphysics on structuralism in the philosophy of science, since much of our discussion will focus on comparativism about quantity, which is a form of structuralism. But unlike the preceding two chapters, this chapter will not be an attack on the structuralist position under discussion. For comparativism is unproblematic to state in postmodal terms; and although we will be discussing some distinctively postmodal concerns about comparativism, the same concerns arise for noncomparativist views of quantity as well.

We will head in several directions, rather than advancing a single thesis. We will study various ways in which a search for attractive fundamental laws impacts the fundamentality-theoretic metaphysics of quantity.

4.1 The problem of quantity

What is the right metaphysics of quantity? What do quantitative theories tell us about the ultimate nature of the world? These questions are especially pressing given the ubiquity of quantitative properties in physics.

Interpreted in the most straightforward and flat-footed way, quantitative theories employ relational predicates of concrete objects and numbers. Thus to say that o is 5 kilograms is to ascribe the two place predicate 'mass-in-kilograms(x, y)' to o and the real number 5. Or, reifying properties and relations, it is to say that o

---

[1] See Wolff (2020) for an account of what makes a property quantitative.

bears the mass-in-kilograms relation to the real number 5. And to say that $F = ma$ (neglecting the directionality of force and acceleration) is to say something like this: 'for any object, x, the number to which x bears the force-in-newtons relation is the product of the numbers to which x bears the mass-in-kilograms and acceleration-in-metres-per-square-second relations'.

(Even though the facts of mass recognized by the flat-footed account involve abstract entities, those facts are capable of a sort of empirical confirmation since they are correlated with observable facts. If x bears the mass-in-kg relation to 5 and y bears it to 4 then x will be more massive than y, which can be observed by placing x and y on a set of scales. The confirmation is admittedly indirect, but no one open to the kinds of metaphysical questions we've been entertaining can insist on direct confirmation of each bit of a foundational theory.)

Does the flat-footed account yield an adequate metaphysics? Your answer will turn on which metaphysical tools you accept.

Suppose the only metaphysical tools you accept are ontological—the only metaphysical questions you recognize concern which entities exist. If you accept the existence of numbers, you might be happy with the flat-footed account.

If you also accept modal tools, you might still be happy. For instance, in modal terms you could recognize necessary connections between numerical predicates corresponding to different scales:

> Necessarily, for any object x and real number y, mass-in-kilograms(x, y) if and only if mass-in-grams$(x, 1000y)$.

You could also recognize necessary connections between the numerical predicates and further, nonnumerical predicates:

> Necessarily, for any objects x, x' and real numbers y, y', if mass-in-kilograms(x, y) and mass-in-kilograms(x', y') then: $y > y'$ if and only if x is more massive than y.

Viewed through a modal-cum-ontological lens, the flat-footed account is still on its feet.

But suppose you accept richer metaphysical tools. Suppose in particular that you accept concept-fundamentality, so that a fundamental theory must specify which concepts are fundamental. You then face the question of which quantitative concepts are fundamental.

The flat-footed account suggests an answer: that relational predicates like 'mass-in-kilograms(x, y)' (assuming that mass is fundamental) express fundamental concepts, or, reifying relations, that mass-in-kilograms is a fundamental relation.

One might object that mass is *physical* and thus couldn't involve abstract entities at the fundamental level. But who made that rule?[2]

A more serious problem arises when we ask *which* relations to numbers are fundamental. In particular, which units are involved? In the case of mass, is the

---

[2] See Sider (2013, pp. 287–8) on this sort of issue. We will return to it in section 4.7.

fundamental relation mass-in-kilograms? mass-in-grams? mass-in-some-other-unit?

It would surely be intolerably arbitrary to say that one of these relations, mass-in-kilograms say, is fundamental to the exclusion of all the others. But the only alternative, apparently, is to say that *all* of the relations are fundamental, which would amount to accepting massive redundancy in the fundamental properties and relations.[3]

This problem, notice, simply doesn't arise if one is only concerned with ontological or modal issues. It arises when one accepts the demand to provide an account of the fundamental nature of mass, by saying which concepts or relations of mass are fundamental. (It also arises if one's metaphysical tool of choice is ground or fact-fundamentality rather than concept fundamentality, if ground and fact-fundamentality are "hyperintensional" enough to distinguish between, for example, the fact that an object is 1,000 g and the fact that it is 1 kg. For then one will still face the question of which numerical units are involved in the fundamental or ungrounded facts of mass.)

Some difficult general questions about fundamentality intrude. *Some* redundancy in the fundamental concepts seems hard to avoid: how to choose between, for instance, parthood and overlap as a fundamental concept for mereology? But then, how much redundancy is tolerable? And the problem of redundancy only becomes worse if, like me, you think that fundamentality can be applied to logical concepts such as existential and universal quantification, or conjunction, disjunction, and negation. We will return to these issues in Chapter 5. For now I continue to assume that massive redundancy in the fundamental concepts is to be avoided, as is arbitrariness, and hence that the flat-footed account does not yield a viable metaphysics of quantity.

4.2 Simple absolutism

We need a different idea about what the fundamental mass concepts are, one that avoids privileging a single unit. Here is one such idea:[4]

Simple Absolutism the "determinate masses" are the only fundamental mass properties or relations.

By the determinate masses, I mean a certain infinite set of properties, each of which is the property of having a certain completely specific mass: the property *having exactly this mass*, the property *having exactly that mass*, and so on. Although we might name such properties by mentioning numbers and a unit ('having exactly

[3] Compare Field (1980; 1985), though his concern is nominalism rather than a thesis about what is fundamental.

[4] Nothing deeper than smoothness of exposition is behind the formulation in terms of properties and relations rather than concepts; recall from section 1.8 that being a fundamental concept is a purely worldly matter.

5 kg mass'), the numbers aren't "built into" the properties—there is no privileged association between the properties and the numbers.

Simple absolutism avoids the problems of the flat-footed view of the previous section. But it goes too far in the opposite direction, away from numbers. The representation of mass using numbers is essential to science. Something fundamental must surely underwrite this procedure; the fundamental facts about mass must constrain the assignment of numbers to massive objects. But if the determinate masses were the only fundamental properties and relations of mass, what would be wrong with assigning to my mass a number that is only slightly smaller than, or even larger than, the mass of a blue whale? We'll make this argument more precise soon.

4.3 Representation theorems and comparativism

Much of the literature on quantity has been in the philosophy of science, where metaphysical concerns are not always central. But the main theories there have close metaphysical cousins, and these cousins can avoid the difficulties faced by the flat-footed account and simple absolutism.

One such cousin may be called, following Dasgupta (2013), 'comparativism'. (Here I'll state a fundamentality-theoretic version, but the label 'comparativism' has more general application, including for example Dasgupta's (2013; 2020) ground-theoretic approach.[5]) According to comparativism, the fundamental quantitative concepts ascribe quantities to concrete objects comparatively, or relationally. In giving a fundamental description of the facts about mass, for example, instead of saying that some object a has mass 1 kg and another object b has mass 2 kg, the comparativist might instead say that b is twice as massive as a. Instead of saying that points p_1 and p_2 are two metres apart and that points p_3 and p_4 are also two metres apart, a comparativist might instead say that p_1 and p_2 are equidistant from p_3 and p_4. Thus fundamental properties and relations specifying the absolute values of quantities are rejected, in favour of fundamental relations which, like *being twice as massive as* and *being equidistant from*, specify the quantitative features of concrete objects relationally.

As David Baker (2020) has pointed out, when comparativists say that the fundamental features of mass (for instance) are relational, they don't have in mind just any relational features. They do not, for example, have in mind relations like the one that holds between x and y just when x is 1 kg and y is 2 kg. For this relation ascribes particular values of mass to particular objects (despite being a relation and not a property); and comparativists think that doubling the mass of all objects in the universe is a distinction that makes no difference at the fundamental level,[6] just

---

[5] See Martens (2017a, pp. 9–10) for extensive references on absolutism and comparativism.

[6] I presume a "four-dimensionalist" approach to time (see Sider (2001) for an overview). Thus the mass relations can hold between objects located at different times, and the envisioned mass-doubling is for objects at all times. A Lewisian (1986a) realist about possible worlds could say that mass relations

as relationalists about space (like Leibniz) think that displacing all objects a fixed amount in one direction is a distinction that makes no fundamental difference. Indeed, comparativism is often embraced precisely because such distinctions are thought not to be genuine.

Rather than seeking a general definition of the kinds of relational features whose embrace as fundamental is consistent with deserving the name 'comparativist', I will proceed in piecemeal fashion: I will consider particular comparativist views, which say that particular chosen relations are fundamental.[7] The main view to be considered we may call simply:

Comparativism about mass There are just two fundamental mass properties or relations: a mass-ordering relation $\succeq$ and a mass-concatenation relation C. $x \succeq y$ holds when x is at least as massive as y; $Cxyz$ holds when x and y's combined masses equal z's.

The glosses on $\succeq$ and C are not meant to be definitions in terms of an underlying numerical scale. For instance, '$x \succeq y$' is not defined as meaning that the real number that is x's mass in grams, say, is greater than or equal to the real number that is y's mass in grams (which would return us to the problems of the flat-footed account). $\succeq$ isn't defined in this way, or in any other way for that matter; it is said to be fundamental, after all. To be sure, numerical scales might facilitate our understanding what the comparativist has in mind by '$\succeq$' and 'C'. Given a practice of measuring mass by numbers, an object x will be assigned at least as large a number as an object y if and only if $x \succeq y$, and an object z will be assigned the sum of the numbers assigned to objects x and y if and only if $Cxyz$ (this is true regardless of the numerical scale—regardless of the chosen units for mass, that is). And since we're familiar with numerical scales, this fact might help us grasp what the comparativist means by '$\succeq$' and 'C'. But as we'll see, the comparativist grounds the practice of measuring mass by numbers in comparative relations like $\succeq$ and C, rather than the other way around.

To dispel lingering worries about the claim that $\succeq$ and C are not defined in terms of numerical scales, it may help to note that there are physical processes that can be used to directly test whether these predicates apply, without knowledge of a numerical scale. A good test for whether $x \succeq y$ is to put x and y on opposite ends of a set of scales and see whether y's side fails to move downwards; and a good test for whether $Cxyz$ is to put x and y on one side and z on the other and see whether they balance.[8]

hold between objects from different possible worlds (although see Lewis (1986a, pp. 76–8)), and thus that a pair of worlds can differ after all by a mere mass doubling, the difference being constituted by mass relations to objects in other possible worlds. This is arguably unavailable to reductionists about possible worlds, but the issue is complex.

[7] See Baker (2020, section 2) for a more general formulation.

[8] These are defeasible: the scales may be defective, the objects may overlap, etc.

The comparativist's fundamental relations, then, do not *presuppose* underlying numerical scales. However, the rich structure of these relations *induces* numerical scales. For given certain assumptions about how $\succeq$ and C behave, one can prove so-called representation and uniqueness theorems:

Representation theorem   There exists at least one function m from individuals to positive real numbers, subject to the constraints (i) $m(x) \geq m(y)$ if and only if $x \succeq y$ and (ii) $m(x) + m(y) = m(z)$ if and only if $Cxyz$.

Uniqueness theorem   Any two functions m and m' obeying constraints (i) and (ii) in the representation theorem are scalar multiples—i.e. for some positive real number k, $m(x) = km'(x)$ for all individuals x.

Similar theorems may be proven for other quantities, provided the comparativist's relations for those quantities—relations analogous to C and $\succeq$—obey appropriate assumptions.

These theorems establish the existence and uniqueness of certain functions, which we may call 'representation functions' in general, and 'mass functions', 'distance functions', 'charge functions', and so on in particular cases. These functions assign numbers to concrete objects that are correlated with the comparativist's fundamental relations for the quantity in question. For example, a mass function assigns at least as high a number to one individual as to another if and only if the first individual is as or more massive than the second (iff, that is, they stand in the $\succeq$ relation). Representation functions are studied in detail in the discipline known as measurement theory.[9]

Representation functions enable us to explain the use of numbers to represent quantities, without requiring the flat-footed metaphysics according to which the fundamental physical facts involve real numbers. Numerical statements about mass may be understood as concerning some chosen mass function, whose values are systematically correlated with the comparativist's nonnumeric fundamental mass facts. Numerical talk about mass is just a way of coding up facts about $\succeq$ and C. The representation theorem guarantees that some mass function or other exists, and the uniqueness theorem shows that the range of mass functions corresponds exactly to the (arbitrary) choice of a unit of measurement.

I complained above that simple absolutism cannot "underwrite" the use of numbers to measure quantities, that it supplies nothing at the fundamental level to constrain the assignment of numbers to masses or massive objects. We can now see what this means: simple absolutism supplies no fundamental properties or relations that can constrain representation functions and enable the proof of representation and uniqueness theorems.

The move to comparativism from the flat-footed mass-in-kilograms view can be regarded as fitting the structuralist template. According to the flat-footed view, the fundamental facts of mass are an array of numerical values distributed over

---

[9] See for instance Krantz et al. (1971a).

individuals. But in this array, too much significance is accorded to the individual nodes, to the individual numbers assigned (this was the arbitrariness objection). All that matters to physics, it might be said, is the relations between the numbers—specifically, their ratios. (One could also make the familiar structuralist epistemic complaint that we can have no knowledge of the particular numerical values.) So we should choose a metaphysics on which only the structure of the array is accorded fundamental significance. Comparativism is such a metaphysics.

Comparativism is an instance of the "replacement" strategy for dealing with problematic entities (section 2.6). The problematic entities were the individual mass values, and they were simply jettisoned, and replaced by the fundamental predicates $\succeq$ and C. (Simple absolutism also jettisoned the numbers, but jettisoned too much since insufficient fundamental structure remained to underwrite the use of numbers to measure quantities.) But the objects to which these replacement predicates apply, namely concrete massive objects, are also embraced at the fundamental level. This particular instance of structuralism is therefore not problematic in the way that the structuralist views considered in Chapters 2 and 3 were. There is no metaphysical funny business like ungrounded existential sentences, for instance.

4.4 Laws and simple absolutism

I've claimed that simple absolutism cannot "underwrite" the use of numbers to measure mass because it supplies no fundamental relations to constrain representation functions. This problem can be sharpened and deepened: simple absolutism would not allow simple laws involving mass. For the only general statements that could govern the simple absolutist's fundamental mass properties would have a list-like, infinitary nature, such as: 'if a particle has exactly *this* mass and experiences exactly *that* force, then it will undergo exactly *such-and-such* acceleration; but if it has exactly this other mass and experiences exactly that other force, then it will undergo exactly thus-and-so acceleration; and if . . .'.

If "laws" of this sort were admissible, there would be no need to posit fundamental properties of charge, mass, force, and so on, or their successors in modern physical theories. In a deterministic context, for instance, instead of a law constraining the motions of particles by their properties, one could instead have a list-like law specifying the allowable trajectories: 'the entire history of particle trajectories could be *thus*, or it could be *so*, or . . .'. It's only our finitude, one might say, that requires us to speak of mass and charge.[10]

According to the simplicity-centric epistemology of fundamentality discussed in section 1.8, we should recognize fundamental concepts that enable the formulation of simple and strong laws. Other things being equal, putative fundamental

---

[10] Compare Melia (1995) on ontological commitment to numbers. This method for eliminating the properties must be distinguished from Allori's (2022a; 2022b), who appeals to a robust conception of laws with a primitive, apparently nonqualitative connection to the objects they govern, and from the ramsification method, on which see section 4.12.

concepts that enable such laws are to be adopted; and a conception of the fundamental concepts that leads to no such laws is to be rejected. This epistemology
speaks against both simple absolutism and the attempt to dispense with fundamental properties altogether: those views do not allow for simple laws of motion.

The use of numbers to measure quantities is not merely a convenient short-
hand, needed only because of our finite nature. Numerical representation is also
intertwined with the existence of simple laws. The statement that an object's
acceleration is directly proportional to the net force on that object and inversely
proportional to its mass is essentially tied to those structural features of mass,
force, and acceleration in virtue of which talk of proportionality is well defined;
to state simple and strong laws, where simplicity is measured by reference to the
fundamental concepts occurring in the law, one must recognize such structural
features at the fundamental level.

This objection to simple absolutism introduces a theme that will occupy us
for the rest of this chapter. If a metaphysical account of quantity is formulated
using the tools of fundamentality, it says something distinctive about the nature
of the fundamental properties and relations. As a result, it may have distinctive
implications about the laws of nature, in which the fundamental physical properties
and relations figure. And given the methodology of fundamentality we have
adopted, this can then inform whether that account should be accepted. Simple
absolutism failed to allow for simple, powerful laws of physics, and thus must be
rejected, whereas comparativism so far appears to be consistent with such laws.
But as we will see, there are a number of questions about the kinds of laws allowed
by comparativism and other accounts of quantity.

4.5 Existence assumptions and intrinsic laws

Another case in which considerations about laws bear on the metaphysics of quantity involves comparativism and Hartry Field's project of formulating "intrinsic"
laws of nature. My chief claim in this section will be this: the main impact on comparativism of the failure of strong "existence assumptions" would be the inability
to formulate intrinsic laws.

By strong existence assumptions I mean certain assumptions about the existence of entities standing in the comparativist's relations, assumptions according
to which there exist infinitely many concrete objects. Here are two representative
examples, in the case of mass:[11]

Existence of sums For any x and y there is some z such that $Cxyz$

Density If $x \succ y$ then for some z, $x \succ z \succ y$,

where '$x \succ y$' means that $x \succeq y$ but $y \not\succeq x$—i.e. that x is more massive than y.

For reasons we will discuss below, it is usually thought that comparativists
must make at least one strong existence assumption. It is also often thought that

---

[11] These are examples of what Krantz et al. (1971b, p. 23) call solvability axioms.

this constitutes a problem for comparativism. The principle of Existence of sums, for instance, implies that there are objects of arbitrarily large mass; why think this is true?

One might defend certain existence assumptions in certain cases. According to Field's comparativist approach to mass, mass is a scalar field on points of space, and on this approach Density is a natural assumption to make, since it is guaranteed to be true provided the field varies continuously. (Existence of sums would not be a natural assumption to make, but Field relies only on Density.[12]) However, it may be objected that gaps in the mass field should not be ruled out, nor should treating mass as a field be mandatory (Arntzenius and Dorr, 2011, p. 227). In any case, the question I would like to ask is this: what would go wrong if the existence assumptions are false? Are comparativists committed to existence assumptions, and if so, why?

The reason that comparativists are usually assumed to be committed to existence assumptions is that the usual proofs of representation and uniqueness theorems in measurement theory make such assumptions. But I think the need for existence assumptions here is not deep.

First, although the usual *proofs* of representation theorems appeal to existence assumptions, the theorems themselves don't depend for their truth on the existence assumptions since a representation theorem will hold provided the actual structure of objects is embeddable within a structure in which the existence assumptions hold.

The truth of uniqueness theorems does indeed depend on existence assumptions. (Which is presumably why measurement theorists freely appeal to existence assumptions in proving representation theorems—they'll be needed eventually anyway. Indeed, a representation and a uniqueness theorem are sometimes combined into a single theorem.) If there are only finitely many massive objects, for example, then it won't in general be true that any two mass functions differ only by a multiplicative constant. (For example, if there are just two massive objects, a and b, where $a \succeq b$ and nothing bears C to anything, then any function that assigns at least as great a number to a as to b counts as a mass function.) But what problem, exactly, does this create? In a world like our own, in which there are a great many massive objects, the various mass functions won't stray *too* far from being scalar multiples (this will become clearer below when we explore the comparativist account of ratios). So there is a sense in which the failure of uniqueness theorems wouldn't much affect the ordinary scientific practice of ascribing quantities to particular objects using numbers. Even if those numerical values aren't unique (even after selecting units), the range of acceptable values might well be within experimental error, so that our customary assumption of uniqueness would be a harmless fiction.

---

[12] See Field (1980) note 41 (and pp. 72–3).

The failure of uniqueness would seem to pose a greater threat to laws, which we assume to hold exactly, regardless of experimental error. Suppose for the moment that a numerical law is some sort of statement about representation functions. (We'll examine this idea in more detail in section 4.9 below.) If existence assumptions fail then there won't be laws with a unique mathematical form: under different choices of representation functions for the various quantities, those functions will stand in different mathematical relationships. Under one triple of functions for mass, force, and acceleration, for example, the magnitude of acceleration might always be proportional to the ratio of force to mass, whereas under another triple that might not be true.

Although this concern is valid, I don't think it's the heart of the problem, since the following comparativist response has some plausibility. Suppose the uniqueness theorems fail, so that different choices of representation functions stand in different numerical relationships. Provided the world is reasonably complex, only one of these numerical relationships will be simple, such as the exact proportionality of acceleration to the ratio of force to mass. Other, more complex relationships could be excluded as laws on that basis alone.

This response, however, assumes that the laws are about representation functions, and thus are "extrinsic" in a certain sense that was introduced by Field. Suppose instead that the laws must be formulated "intrinsically", as Field insists they should be. Then, as we will see, there is an intractable problem caused by the failure of uniqueness assumptions. This, I think, is the heart of the matter.

In *Science without Numbers*, Field defended a conception of physical theories based on a comparativist metaphysics of quantities.[13] The main point was to defend nominalism: physics was to be rewritten in terms of predicates like 'C' and '$\succeq$', which relate concrete objects rather than numbers. But a secondary point was to enable the formulation of what Field called intrinsic laws and explanations. Even those who believe in the existence of numbers and other abstract entities, Field thought, should reject the idea that laws of nature and physical explanations make reference to them. Physical explanations ought to characterize their subject matter "directly", and not indirectly via abstract objects.

A statement about representation functions is a paradigm of what Field means by a non-intrinsic, or extrinsic, statement. Such statements would characterize the fundamental comparative predicates only by their relationships to representation functions. Laws of this sort concerning mass, for example, would be saying 'the facts about $\succeq$ and C are such that, when they are coded up using mathematical entities in a certain way, the codes stand in certain simple mathematical relationships'. This feels artificial, indirect, and distant from the subject matter of physics.[14] We

---

[13] Though not of the fundamentality-theoretic flavour, not explicitly anyway.

[14] Somewhat similar concerns have been expressed about other intrusions of abstracta into intuitively alien subject matters. It's odd, for instance, to think that speaking of ancestors commits one to classes (Boolos, 1985, p. 327).

should instead seek intrinsic laws, simple statements that directly concern the comparative predicates. To be sure, such statements would, given representation and uniqueness theorems, have extrinsic statements as consequences, but those extrinsic statements would not be the true laws, and should not be cited in the best explanations.

Arntzenius and Dorr (2011) agree with the need for intrinsic laws, and call the search for them the 'hard problem' of quantity. The requirement that laws be intrinsic has great initial appeal—especially for a realist about concept-fundamentality. In section 4.7 I will question it. But for now my point is just that it requires strong existence assumptions.

I don't have a general argument that comparativist intrinsic laws need existence assumptions; but this is clearly true given the best-developed approach to intrinsic laws, namely Field's. Field's method uses quantification over "standard sequences" as described in Krantz et al. (1971a). A standard sequence for a quantity Q—mass, say—is a sequence of massive objects that is "evenly spaced" in the sense that the difference in mass between adjacent objects in the sequence is constant:

$$1q \qquad 2q \qquad 3q \qquad 4q \qquad 5q \qquad 6q$$

(The sequence can be arbitrarily long, or even infinite.) Think of a standard sequence for quantity Q—a "Q sequence"—as a "grid" one can lay down on objects. Most objects do not have exactly the same Q value as a point of the grid (only those whose Q values are integer multiples of the grid's unit—the Q value of its first member—do), but they can nevertheless be represented as points on the grid with accuracy that increases as the grid's resolution increases—that is, as the size of its unit decreases. By quantifying over Q-sequences of increasingly high resolution Field develops intrinsic correlates of statements about, for example, ratios between real-valued quantities. And the problem is that if Density fails then this strategy fails since there won't exist arbitrarily fine-grained grids.

In more detail: let Q_1 and Q_2 be two quantities that are ratio scales[15] (ratios of values of these quantities are significant; mass is an example), with corresponding fundamental relations $\succeq_{q_1}, C_{q_1}, \succeq_{q_2},$ and C_{q_2}; and let q_1 and q_2 be representation functions for Q_1 and Q_2, respectively. Suppose we want an intrinsic statement of this:

$$(*) \quad \frac{q_1(x)}{q_1(v)} < \frac{q_2(u)}{q_2(y)},$$

where $x, y, u,$ and v are any objects. Here is a simplified method, in the spirit of Field's. We'll be using mereology; and for simplicity, let's assume atomism and also that all objects possessing Q_1 and Q_2 are mereological atoms. (This assumption

---

[15] Field's method also applies to other quantities, e.g. quantities where it is ratios of differences that are significant.

could be avoided, but it's harmless in Field's case anyway, since the quantities in question are taken to be fields defined at points of space.) First a definition, for any ratio scale quantity with corresponding $\succeq$ and C:

> S is a Q *sequence* if and only if S has an atomic part, s_1, such that for every atomic part x of S (except perhaps a final part, which bears $\succeq$ to every other atomic part of S), there is some atomic part y of S such that: (a) Cs_1xy, and (b) for any atomic part z of S, if $y \succeq z \succeq x$ then $z = x$ or $z = y$.

'Sequence' is now technically inaccurate: Q sequences are mereological sums, not set-theoretic sequences. A Q sequence contains as parts (not members) some atoms $s_1, s_2, s_3, \ldots$ where $q(s_k) = kq(s_1)$ (for any representation function q for Q), and thus can (in part) be pictured as in the diagram above. An intrinsic correlate of (*) can then be given:

> (**) There exists a Q_1 sequence S_1 and a Q_2 sequence S_2 such that (i) x is an atomic part of S_1; (ii) y is an atomic part of S_2; (iii) there are exactly as many atomic parts of S_1 that are $\preceq_{q_1} x$ as there are atomic parts of S_2 that are $\preceq_{q_2} u$; and (iv) there are fewer atomic parts of S_2 that are $\preceq_{q_2} y$ than there are atomic parts of S_1 that are $\preceq_{q_1} v$.

($\preceq$ is just the converse of $\succeq$. Note the use of the generalized quantifiers 'there are exactly as many Fs as Gs' and 'there are fewer Fs than Gs'; see Field (1980, chapter 9) for discussion.) (**) is an intrinsic correlate of (*) in the sense that—and this can be proven, given certain assumptions; compare Field (1980, note 48)—for any representation functions q_1 and q_2 for Q_1 and Q_2, (**) holds if and only if (*) holds under q_1 and q_2.

But this correspondence between (*) and (**) depends on Density. As mentioned earlier, we can think of Q sequences as grids for measuring Q values; the correspondence relies on the ability to choose grids of arbitrarily high resolution. If Density fails then the grids won't be guaranteed to exist, and as a result, even if (*) "ought" to be true, (**) might nevertheless be false: there might not exist the appropriate sequences S_1 and S_2.

This then threatens the existence of intrinsic laws concerning ratios of quantities. To illustrate, consider Newton's second law $F = ma$. To keep things simple, pretend that acceleration is a primitive scalar quantity taking only positive values (so that it's a ratio scale), and pretend that all objects undergo exactly the same net force. Under these assumptions, Field would take the law—call it 'Simple-Newton'—to say that the product of mass and acceleration for any object is the same as for any other object. That is, for any x and y:

$$m(x)a(x) = m(y)a(y)$$

or, equivalently:

$$\frac{m(x)}{m(y)} = \frac{a(y)}{a(x)}$$

or, equivalently:

$$\frac{m(x)}{m(y)} \not< \frac{a(y)}{a(x)} \text{ and } \frac{a(y)}{a(x)} \not< \frac{m(x)}{m(y)}.$$

Applying (**), this has the following Fieldian intrinsic correlate:

I-Newton For any objects x and y: there do *not* exist a mass sequence S_1 and an acceleration sequence S_2 such that (i) x is an atomic part of S_1; (ii) x is an atomic part of S_2; (iii) there are exactly as many atomic parts of S_1 that are $\preceq_m x$ as there are atomic parts of S_2 that are $\preceq_a y$; and (iv) there are fewer atomic parts of S_2 that are $\preceq_a x$ than there are atomic parts of S_1 that are $\preceq_m y$; and there do *not* exist an acceleration sequence S_1 and a mass sequence S_2 such that (i) y is an atomic part of S_1; (ii) y is an atomic part of S_2; (iii) there are exactly as many atomic parts of S_1 that are $\preceq_a y$ as there are atomic parts of S_2 that are $\preceq_m$ than x; and (iv) there are fewer atomic parts of S_2 that are $\preceq_m y$ than there are atomic parts of S_1 that are $\preceq_a x$.

Suppose, now, that Density and other strong existence assumptions are false. I-Newton could then be true simply because of the nonexistence of appropriate mass or acceleration sequences. And if I-Newton can so easily be true, that means that it doesn't have the consequences we would expect it to have in those circumstances. In particular, it won't have the consequences about C_m, C_a, $\succeq_m$, and $\succeq_a$ that the numerically stated law Simple-Newton has. For example, Simple-Newton implies that there cannot be a pair of objects standing in both of the defined relations $\succ_m$ and $\succ_a$ in the same order. (If $v \succ_m u$ and $v \succ_a u$ then for any representation functions m and a for mass and acceleration, $m(v) > m(u)$ and $a(v) > a(u)$, and so $\frac{m(v)}{m(u)} > 1$ and $\frac{a(u)}{a(v)} < 1$, and so $\frac{m(v)}{m(u)} \neq \frac{a(u)}{a(v)}$, violating Simple-Newton.) But I-Newton doesn't have this consequence without assuming Density or some other strong existence assumption.[16] It therefore cannot be regarded as a *law* (even though it's true) in such circumstances: a law must be an appropriately strong statement.

The point has been illustrated with an extremely simplified example, but it holds, I take it, generally: strong existence assumptions are needed for Fieldian comparativist[17] intrinsic laws, since they quantify over standard sequences. If

[16] Consider a model in which u and v are the sole massive and accelerated elements in the domain. (The model must also be a model of mereology; so let u and v be mereological atoms; and let the model also contain another object to be the fusion of u and v, which is not in the field of the mass and acceleration relations). Let there be no cases of either C_m or C_a; let $v \succ_m u$ and $v \succ_a u$; let $\succeq_m$ and $\succeq_a$ be reflexive. Other than existence assumptions, the usual axioms for extensive systems hold for both mass and acceleration. For example, convert the axioms for an extensive structure with no essential maximum in Krantz et al. (1971a, vol. 1, p. 84) to the present notation by defining $Cxyz$ to mean their '$(x,y) \in B$ and $z \sim x \circ y$', and then remove the assumption of Density (their axiom 4). The resulting axioms then hold in the model, for $\succeq_m$ and C_m as well as for $\succeq_a$ and C_a. And it's easy to verify that I-Newton also holds: the only mass- or acceleration-sequences are the "one-membered" sequences u and v, which makes both halves of I-Newton trivially true.

[17] I have been understanding 'comparativism' narrowly, as the view that C and $\succeq$ are the fundamental mass relations. But on a broader construal other views would be included. Bigelow and Pargetter

the quantification is "negative" (as it is in I-Newton), saying that there do *not* exist certain standard sequences, then the Fieldian intrinsic statement will be too weak to be a law, if the existence assumptions fail. If, on the other hand, the quantification is positive (as in (**)), saying that there *do* exist certain standard sequences, then the Fieldian statement may fail to be true even when it shouldn't, if appropriate standard sequences don't exist because of the failure of existence assumptions.

4.6 Intrinsic laws and Mundy's multigrade view

For another illustration of the importance of intrinsic laws to the metaphysics of quantity, consider Brent Mundy's (1989) ingenious attempt to avoid problematic existence assumptions in the comparativist theory of quantity. Mundy's strategy was to combine and enhance the standard predicates $\succeq$ and C into a single primitive multigrade predicate '$a_1, \ldots, a_n \succeq b_1, \ldots, b_m$', meaning, intuitively, that $a_1, \ldots, a_n$ together have a sum total of the quantity in question that is at least as great as the sum total of the quantity possessed by $b_1, \ldots, b_m$. The predicate is "multigrade" in the sense that it does not have a fixed number of argument places—any finite number of arguments can be included on its left- and right-hand sides.

The main virtue of this approach can be seen as follows. Suppose we want to say that the amounts of the quantity in question possessed by a pair of objects x and y stand in a ratio that is at least as great as some particular fraction $\frac{n}{m}$. Given standard comparativism we can do this with standard sequences, in one of two ways. We can speak of standard sequences in which x and y are the terminal members, "finitely dividing" x and y's portion of the quantity as finely as needed:

There exist $x_1, x_2, \ldots, x_{n-1}$ and $y_1, y_2, \ldots, y_{m-1}$ such that (i) $Cx_1 x_1 x_2$, $Cx_1 x_2 x_3$, ..., $Cx_1 x_{n-1} x$; (ii) $Cy_1 y_1 y_2$, $Cy_1 y_2 y_3$, ..., $Cy_1 y_{m-1} y$; and (iii) $x_1 \succeq y_1$

thus relying on Density. Or we can speak of standard sequences in which x and y are the initial members, "finitely copying" x and y as many times as needed:

There exist $x_1, x_2, \ldots, x_{m-1}$ and $y_1, y_2, \ldots, y_{n-1}$ such that (i) $Cxxx_1$, $Cxx_1 x_2$, ..., $Cxx_{m-2} x_{m-1}$; (ii) $Cyyy_1$, $Cyy_1 y_2$, ..., $Cyy_{m-2} y_{m-1}$; and (iii) $x_{m-1} \succeq y_{n-1}$

(1988), for instance, posit a domain of comparative relations over individuals (twice-as-massive-as, three-times-as-massive-as, and so forth), and also some higher-level relations that give the first-order relations their structure. This counts as comparativist in a perfectly good sense (global mass-doubling is a distinction without a difference), yet would presumably not need strong existence assumptions about individuals to formulate laws in Field's way: the standard sequences could be constructed from the first-order relations. Thanks to Shamik Dasgupta for this point.

thus relying on the Existence of sums. But on Mundy's multigrade approach, we needn't rely on any existence assumptions; we can just say:

$$\underbrace{x,\ldots,x}_{m \text{ occurrences}} \succeq \underbrace{y,\ldots,y}_{n \text{ occurrences}}.$$

As a result, as Mundy shows, if one uses his multigrade predicate, existence assumptions (like Density and Existence of sums) aren't needed to prove representation and uniqueness theorems.

But when it comes to stating intrinsic laws (which Mundy doesn't discuss), the need for the existence assumptions reappears. Mundy's representation and uniqueness theorems tell us that the totality of facts stateable in his language fix a unique—up to scalar transformation—numerical representation; they don't tell us that a simple law stated in terms of the numerical representation has a single, simple corresponding sentence in his language. Mundy's method for saying 'the mass-ratio between x and y exceeds $\frac{n}{m}$' as described in the previous paragraph is for *fixed* n and m. But in order to construct intrinsic correlates of statements about ratios between real-valued quantities, one needs such comparisons when n and m are variables, as in 'for any integers n and m, if the mass-ratio of x to y is greater than or equal to $\frac{n}{m}$ then the mass-ratio of z to w is greater than or equal to $\frac{n}{m}$'. Field's method for doing this quantifies over standard sequences and thus requires existence assumptions. In essence it is a modification of the method of division to the case where n and m are variables; a standard sequence is a single entity corresponding to the sequence of values of the existential quantifiers in the method of division (and also, the quantification over natural numbers is avoided by the use of generalized quantifiers). Mundy's method cannot be modified in this way, for without making existence assumptions we have no entities corresponding to his sequences of occurrences of a single variable.

4.7 Intrinsicality of laws

I have said that the main problem that would be caused by the failure of the existence assumptions would be the inability to state intrinsic laws. But why think that laws should be intrinsic?

As we saw in section 1.8, once the notion of fundamental concepts is assumed, it is natural to take a general bias in favour of simplicity in theory choice to require, at least in part, a bias in favour of simple laws as formulated using the fundamental concepts. But this assumption doesn't itself require intrinsic laws, since even extrinsic laws are simple statements about the fundamental concepts, if the comparativist's concepts are fundamental (and if the mathematical concepts are fundamental).

Field himself said three main things against extrinsic laws. He complained about formulations of laws that 'appeal to *extraneous, causally irrelevant* entities' (Field, 1980, p. 43, my emphasis), and he complained about *arbitrariness*. Let's take these individually.

4.7.1 *Causal irrelevance*

Field expands on the charge of causal irrelevance:

If, as at first blush appears to be the case, we need to invoke some real numbers like 6.67×10^{-11} (the gravitational constant in $m^3/kg^{-1}/s^{-2}$) in our explanation of why the moon follows the path that it does, it isn't because we think that that real number plays a role as a cause of the moon's moving that way.

(Field, 1980, p. 43)

But in what sense would fundamental extrinsic laws render real numbers "causally relevant"? It wouldn't imply, for example, that we can see or touch real numbers, or that purely numeric facts about real numbers (such as that $3 = 2 + 1$) cause or are caused by physical facts such as that I am sitting, am in a certain location, am more massive than my cat, etc. In what sense would recognizing fundamental extrinsic laws require us to grant the real number 6.67×10^{-11} "a role as a cause of" the moon's motion? It wouldn't imply that the fact that the number exists, or the fact that it has certain mathematical properties, causes the moon to move as it does, or that different possibilities for the moon's motion correspond to different possibilities for the purely mathematical facts about this and other real numbers.[18] The involvement of real numbers in causes of (or causal explanations of) the moon's motion is only in mixed physical-cum-mathematical facts like the law of gravitation. And if the complaint is that numbers aren't causally relevant even in that sense, the complaint is merely the insistence that numbers aren't involved in fundamental laws. But again, who made that rule?

4.7.2 *Extraneous entities*

Consider next the claim that a fundamental law should involve no "extraneous" entities. We have, it would seem, a conception of the proper subject of certain laws, of the kinds of entities those laws really concern. Other entities are extraneous, and the laws should not name or quantify over such entities. Real numbers and functions, for instance, seem not to be part of the proper subject of Newton's second law; the law should concern no entities other than points of space or space-time or bodies in motion.

This is a more promising complaint, but it's still far from clear. What exactly makes an entity "extraneous"? As Joseph Melia (1998, section 2) has pointed out, even Field's preferred laws appear to involve entities that are in some sense extraneous—entities that seem intuitively not to be involved in the facts the law governs. The Fieldian intrinsic correlate of the claim that x and y share the same product of mass and acceleration quantifies over mass and acceleration sequences containing x and y. Such sequences contain arbitrarily many massive and accelerated objects, which can be located anywhere whatsoever; the sequences

---

[18] It might be objected that the existence of 6.67×10^{-11} *is* a cause of the moon's moving as it does, given the extrinsic law, since if the number hadn't existed then the extrinsic law wouldn't have been true, in which case the moon would have moved differently. But why should the comparativist admit the inference to the final claim?

are, intuitively, no less extraneous than real numbers. To be sure, the sequences are composed of bodies in motion. And, being mereological fusions rather than set-theoretic sequences, they are perhaps not of the wrong "ontological category" in the way that real numbers are. But they nevertheless seem extraneous in Field's sense. Just like real numbers, they are not themselves the proper subject matter of the laws of motion, but rather are being used to "code up" relations between bodies in motion, which are the proper subject matter. So now one worries that comparativism does not enable truly intrinsic laws after all. Even given Field's method for avoiding the extrinsicality of quantification over real numbers, we still have the extrinsicality of quantification over standard sequences.

There is a further (though related) complaint one can make about Field's allegedly intrinsic laws: they are nonlocal in a certain sense.[19] In two senses, actually. First, I-Newton concerns two objects at a time. Second, there is what we already discussed: the quantification over mass and acceleration sequences, where the parts of these sequences can be at arbitrarily large spatial distances from the given pair of objects, and can be anywhere in time. Now, gravitation itself is already nonlocal in Newton's theory. But the kind of nonlocality resulting from quantification over standard sequences of individuals is of a new and distinctive sort, and moreover would be present in Fieldian formulations of local theories such as classical electromagnetism.

(Notice, by the way, the convergence of the discussions of quantity in the philosophy of science and pure metaphysics. A number of metaphysicians have objected that comparativism makes mass relational rather than intrinsic.[20] It is satisfying to see how a brute appeal to "intuition" can be replaced with an appeal to a constraint on laws (though appeal to intuition perhaps returns with the prohibition of extraneous entities). And notice again how important lawhood is for figuring out what is fundamental.)

Thus it is unclear whether Field's intrinsic laws avoid "extraneous" entities any better than laws about representation functions do.

---

[19] To be sure, there remains a sense in which they might still be local: the Fieldian laws might imply a statement about representation functions to the effect that the numerical value of a certain quantity is determined by the numerical values of certain other quantities in the immediate spatiotemporal vicinity. Thanks to Eddy Chen here.

[20] Metaphysicians have also focused on modal considerations: comparativism precludes the possibility of, e.g., everything doubling in mass. Reliance on the "intuition" that this is genuinely possible seems to me to suffer from the same problem as relying on the intuition that mass properties are intrinsic: in each case the intuition is nothing but a belief based on an internalized commonsensical proto-theory of mass that has no independent justification. We ordinarily think of quantities as fundamentally intrinsic, and we then apply combinatorial reasoning to yield the possibility of doubling. One could attempt to give a stronger modal argument by arguing that comparativism allows doubling *one* thing's mass (or anyway allows the comparativist equivalent of this) and then arguing that no reasonable conception of the state-space can allow this but not allow the doubling of everything's mass. But this argument is no good, since a perfectly reasonable conception of the state-space can be given in "native" terms: the state-space consists of all the possibilities, given in some combinatorial way, for the comparativist's fundamental relations. What is possible in other terms can then be "read off" from this.

4.7.3 Interlude: mixed absolutism

In fact, there is a real question of whether *any* reasonable view can avoid extraneous entities. Consider, for example, one of the main rivals to comparativism. Because of the problem of existence assumptions, Brent Mundy (1987) argued that comparativism should be replaced by a view which I'll call 'mixed absolutism':

Mixed absolutism   The only fundamental mass properties or relations are the determinate masses, plus two higher-order "structuring relations" $\geqslant$ and $*$ over the determinate masses. $p \geqslant q$ holds when p is at least as "large" as q; $*(p, q, r)$ holds when p and q "sum to" r.

(He made this proposal before proposing the view discussed in section 4.6.) The determinate masses, recall, are completely specific monadic properties of having a certain mass, with no "built-in" numbers. The structuring relations $\geqslant$ and $*$ are higher-order analogues of $\succeq$ and C. As with $\succeq$ and C, the glosses on $\geqslant$ and $*$ aren't definitions; those relations are fundamental, and not defined in terms of an underlying numerical scale. The view is "absolutist" because it retains the monadic determinate masses. But it is "mixed" because it also includes some fundamental comparative relations, the structuring relations $*$ and $\geqslant$, although they relate determinate masses rather than concrete massive objects. The structuring relations can be used to constrain representation functions and thus to "underwrite" the use of numbers to measure quantities. Representation functions now assign numbers to the determinate mass properties rather than to concrete massive objects. The versions of Density and/or Existence of sums needed for representation and uniqueness theorems say that the set of determinate mass properties, rather than the set of massive objects, is dense and/or closed under sums, which are apparently more plausible assumptions (provided one is happy with an ontology of properties in the first place!).

(Arntzenius and Dorr (2011) defend a related view.[21] They too apply structuring relations (like $\geqslant$ and $*$) to entities other than familiar concreta, which are hypothesized to exist even when there are gaps in which quantities are possessed and hence which can safely be assumed to obey the existence assumptions. But for them, the hypothesized additional entities are not properties, but rather are points in substantival "quality spaces". In the case of mass, for instance, they posit the existence of a "mass space" consisting of points structured by the likes of $\geqslant$ and $*$, as well as a distinguished relation of occupation between massive objects and points in this space. They de-emphasize the importance of whether points in this space should be considered properties in the traditional sense, and stress the analogies between the reasons for positing their quality spaces and the reasons for positing the points of familiar space (pp. 229–30)—rightly in each case, in my view.)

---

[21] See also Wolff (2020, chapter 7).

So: could a mixed absolutist state intrinsic laws that avoid "extraneous entities"? Even a mixed absolutist will still need to quantify over standard sequences, when attempting to construct intrinsic laws concerning ratios. These will no longer be sequences of concrete objects like massive particles, but rather will be sequences of properties (or sequences of points in quality spaces, for Arntzenius and Dorr). But isn't it still true that arbitrary sequences of such properties are, intuitively, "extraneous" to a law governing the mass and acceleration of a given particle? The properties in the sequences are not possessed by the particle, after all. Why would it help that the members of the sequences are properties (or points in quality spaces) rather than concrete things?

Incidentally, mixed absolutism brings out another contrast between the modal and postmodal approaches to metaphysics. Despite the fact that the higher-order relations ⩾ and ∗ presumably hold necessarily whenever they hold, and so are supervenient on anything, as Maya Eddon (2017) has pointed out there is nevertheless a strong case for their being fundamental. The strongest part of that case, to my mind, is that these relations are needed for simple and strong laws; without them we are left with Simple absolutism, which only enables infinitary list-like "laws" (section 4.4). This is another illustration of the shortcoming of modal tools like supervenience as "measures of metaphysical commitment". Supervenient features are not a metaphysical free lunch if they are needed to state laws of nature.

It may be tempting to think that the higher-order relations ⩾ and ∗ are a free lunch, not because they are supervenient, but rather because they are *internal*, holding in virtue of the intrinsic features of their relata.[22] Then maybe simple absolutism is acceptable after all. Although ⩾ and ∗ aren't fundamental, since they are internal they are implicit in something at the fundamental level that the simple absolutist already accepts, namely the determinate mass relations. So perhaps they are available to figure in laws for this reason.

Now, it isn't clear that those relations are internal, since it isn't clear that their relata—the determinate mass properties—have any nontrivial intrinsic properties at all (compare Lewis (1986a, pp. 177–9)). And even if they are internal, it isn't clear that facts about their instantiation are "implicit in something at the fundamental level that the simple absolutist already accepts", since simple absolutism as I defined it does not posit any fundamental monadic properties of the determinate mass properties that would constitute their intrinsic features.

But set all that aside. The real problem with the proposed defence of simple absolutism is that it simply abandons our core epistemic constraint on fundamentality, that we should posit fundamental properties that are needed to formulate simple and strong laws. And actually it's worse than that. The "normal" higher-order structuring relations ⩾ and ∗ wouldn't be the only internal relations over the determinate mass properties. As Eddon (2017, pp. 97–8) points out, there will

[22] Thanks to Kris McDaniel for helpful discussion here.

also be various "gerrymandered" relations $\geqslant'$ and $*'$, which would satisfy axioms of the sort Mundy lays down on $\geqslant$ and $*$, but which would have very different "mass functions" (not scalar transformations of the mass functions based on $\geqslant$ and $*$). Given the Humean view of laws, Eddon shows that there would be no basis for privileging the right laws, based on the genuine mass- and other functions, over the wrong ones based on the gerrymandered relations. Nor would antire-ductionism about laws help. If there are primitive laws about $\geqslant$ and $*$ rather than $\geqslant'$ and $*'$, then the former rather than the latter are functioning as fundamental relations in a nomic sense, anyway; and to my mind the pressure would then be on to treat them as fundamental full stop.

4.7.4 *Arbitrariness*

Back to Field's complaint about extrinsicality. In addition to complaining about causally irrelevant and extraneous entities, Field also says that 'one of the things that gives plausibility to the idea that extrinsic explanations are unsatisfactory if taken as *ultimate* explanation is that the functions invoked in many extrinsic explanations are so arbitrary' (p. 45). For example, laws concerning one particular set of representation functions involve arbitrary choices of unit for each of the quantities involved.

Extrinsic laws that *quantify* over representation functions, rather than con-cerning particular ones, would not turn on arbitrary decisions of this sort. (We will discuss such laws in section 4.9.) Thus the arbitrariness complaint targets only certain sorts of laws involving representation functions.

Further, the complaint would be avoided by a view that involves real numbers even more deeply in the metaphysics of quantity. Like the flat-footed approach to quantities with which we began our discussion, this view characterizes facts about quantities using relations to real numbers, but it avoids privileging a unit by assigning *ratios* of quantities to *pairs* of objects rather than values of quantities to individual objects. The fundamental concept of mass, on this approach, is a (functional) three-place predicate $Mxyr$ applied to two massive things x and y and the real number r that is the ratio of x's mass to y's mass; mass representation functions can then be defined as functions m such that $\frac{m(x)}{m(y)} = r$ if and only if $Mxyr$. This privileges no unit since mass ratios are invariant across units.[23] If the extrinsicality complaint were simply a complaint about arbitrariness, then this metaphysics should be exempt.

Perhaps the arbitrariness complaint can still be made against the ratio view, if real numbers are regarded as being constructed from sets. There are many ways this construction can go. The usual strategy is to construct integers from sets, rational numbers as equivalence classes of pairs of integers, and real numbers as sets of rationals. But at each stage there are arbitrary-seeming choices about how exactly to carry out the construction. If the fundamental mass concept is a relation

---

[23] Thanks to Earl Conee for discussion here; and see Mundy (1988) on ratio spaces.

to real numbers under one chosen construction, this would seem objectionably arbitrary; it's as arbitrary to privilege a method for constructing real numbers from sets as it is to privilege a unit of measurement.[24] So the fundamental facts, for the ratio-relation-to-real-numbers approach, involve an arbitrary element. Similarly, it is arbitrary what to count as representation-function-theoretic laws, given the arbitrariness in how to define real numbers—and functions, and other such mathematical concepts—in fundamental terms. These concerns could be avoided by taking real numbers (and functions, etc.) to be sui generis, but at the cost of inflationary philosophy of mathematics.

But it isn't clear that Field's own approach is exempt from *this* sort of arbitrariness. Field's intrinsic laws depend on a certain method for nominalizing statements about ratios (via standard sequences), and it's hard to believe that this method is the only one that would do the trick. This arbitrariness seems on a par with the arbitrariness in representation-function-theoretic laws corresponding to the arbitrariness in how to construct real numbers from sets. (The arbitrariness in the ratios view is perhaps deeper, since it infects the fundamental facts, not just the laws.)

My overall concern, zooming out: what Field is recoiling from, when he complains about extrinsic laws, once we set aside the complaint about causal relevance, is some combination of arbitrariness and artificiality in the use of numbers to code up a constraint on $\succeq$ and C. That complaint really does mesh with a fundamentality-centric approach to metaphysics. The intuitive idea of the fundamentality approach's epistemology is that the laws ought to look attractive as laws when you view them as they fundamentally are—when you look at their formulation in a completely fundamental language. And what I'm worried about is that laws that look good thus viewed may just not be in the cards. Because of parsimony—a separate constraint on our epistemology—we are drawn to a minimal basis. But that inevitably means that any statement of powerful quantitative laws is bound to involve some artificiality or arbitrariness or both.

If there is no escape from artificiality and arbitrariness, how should we react? Should we say that intrinsicality comes in degrees and that good laws are as intrinsic as possible? Should we try to make a distinction between problematic and unproblematic ways of being extraneous? Neither seems promising. A sceptic about our entire approach to metaphysics through the lens of fundamentality based on a laws-centric epistemology might well be gloating at this point. We will return to this issue.

---

[24] See Sider (1996) for more on this sort of argument. Note that the argument is essentially fundamentality-theoretic.

4.8 Baker's escape velocity argument

Yet another case in which laws are central to the metaphysics of quantity is Baker's (2020) argument that comparativism can undermine determinism.[25] As we will see, there is a comparativist response based on a distinctive—and to my mind difficult to stomach, though perhaps unavoidable—approach to lawhood.

Determinism says roughly that the state of the world at a time determines the state of the world at other times. Whether this is true depends in part on how rich the state of the world at a time is. Since comparativists think that the state of the world at a time is less rich than absolutists and standard physics take it to be, it is certainly a priori possible that deterministic laws could be formulated by reference to the richer absolutist structure, but not by reference to the poorer comparativist structure. (To take a silly example, imagine that particle mass varied continuously, and that particles whose mass reaches a certain absolute threshold instantaneously changed their charge. If one could only refer to mass comparisons and not the absolute threshold, one could not state a deterministic law governing the charge transitions.) What Baker argues is that this a priori possibility is in fact realized in the case of comparativism about mass and Newton's theory of gravitation and motion.

Newton's theory includes two main laws. (Insofar as we're not inclined to reify force, the laws could be combined; but let's keep them separate here.) One law says how objects move, depending on the forces acting on them:

Dynamics Any body will accelerate in the direction of the net force on it, with a magnitude that is the ratio of the force's magnitude and the body's mass:

$$\vec{a} = \frac{\vec{F}_{net}}{m} \qquad \text{that is:} \quad \vec{F}_{net} = m\vec{a}.$$

(The net force on any given body is the vector sum of all the component forces acting on it.) A second law specifies the gravitational component forces—the only forces, in this theory—acting on a given body:

Law of gravitation Each of any pair of bodies exerts a component gravitational force on the other directed towards itself, whose magnitude is the ratio of the product of their masses and the square of the distance between them:

$$\vec{F}_{12} = \frac{Gm_1 m_2}{r^2} \hat{r}_{12}$$

where $\vec{F}_{12}$ is the component force exerted on object 1 by object 2, $\hat{r}_{12}$ is a vector of length one in the direction from object 1 towards object 2, and G is a constant whose numerical value depends on the particular units in which mass, distance, force, and time are measured. Actually, the dynamical law also has a constant, but

[25] See Martens (2017a) for an extended discussion of the argument and related issues.

the standard units of kilograms, metres, seconds, and Newtons are chosen so that its numerical value is 1.

These laws are often said to be deterministic: given them, the past fixes the future.[26] That's roughly because of the availability of the following procedure for determining the future position and velocity of any particle, as a function of the current positions and velocities of all particles (particle masses do not change in this theory): use the law of gravitation plus the masses and positions of all particles to calculate the component forces acting on the particle; then vector-add these to get the net force acting on the particle; then use the dynamics to calculate the acceleration of the particle; then use the particle's acceleration, velocity, and position to determine its position and velocity at the next instant (calculus gets rid of 'next instant'); repeat for each particle.

But Baker argues that the Newtonian laws would *not* be deterministic if comparativism were true. Those laws imply a derived law specifying the *escape velocity* for a given massive body:

$$v_e = \sqrt{\frac{2GM}{r}}.$$

Here M is the mass of the body in question, r is the radius of the body, and v_e is the speed that you would need to give to a projectile located at the body's surface (in a direction away from the common centre of mass of the body plus projectile) in order for that projectile to "escape" the body's gravitational attraction, meaning that if it is never acted on by any other force, the body's gravitational force would never turn the projectile around and bring it back. Notice that in the escape velocity formula there is no dependence on the mass of the projectile, only on the mass of the body. That means that whether the projectile will eventually escape does not depend solely on the quantitative facts about the present that the comparativist recognizes—namely, ratios of quantities—but also depends on a fact that the comparativist does *not* recognize, namely the body's absolute mass. Thus the present, given comparativism, is insufficiently rich to fix the future, since it cannot fix whether the projectile eventually escapes.

To present this argument more carefully, we'll need a more exact definition of 'determinism'. According to a typical statement, it is *laws* that are or are not deterministic; and laws are deterministic when they, together with the state of the world at any time, yield the state of the world at any other time:[27]

Determinism Some laws are deterministic if and only if any two possible worlds in which those laws are true that share their world state at any time share their world state at all times.

---

[26] In fact they aren't quite deterministic, for subtle reasons that we can ignore here. See Hoefer (2016, section 4.1) for an overview.

[27] This is two-way determinism, future-to-past as well as past-to-future. Attention could be restricted to one direction only, but there is no need in the present case since the laws of Newtonian gravitational theory are time-reversal-symmetric.

The notion of "the state of the world" at a time needs to be clarified. The state of the world at a time had better not include *all* features of the world then, since one such feature is *being such that the state of the world at a certain future time is such-and-such*—determinism would be trivially true. So it's customary to restrict which features are included. We might restrict to certain quantities specified by list: mass, position, and velocity at the time in question, say. Or we might restrict using some metaphysical concept: to the *intrinsic*, say, or to the *neighbourhood* (in the sense of Arntzenius (2000)) features of the world at that time.

Consider now a world in which Newton's laws are true and which contains nothing but a planet, Earth, plus a projectile that is launched from the Earth's surface at a speed slightly faster than the Earth's escape velocity. Given the escape velocity law, the projectile will never return. Now consider a second world in which Newton's laws are true that is just like the first one at the initial time, but in which the masses of all objects are doubled. Call the planet in this world Pandora. Pandora's mass is double that of Earth's, and so its escape velocity is higher by a factor of $\sqrt{2}$, and so the projectile in this second world *will* return.

So we have two worlds, one in which the projectile escapes the planet and another in which it does not. These worlds eventually have different world states, even from a comparativist point of view, since the spatial relations in the future are clearly different. But at the initial moment, the worlds would seem to have the same comparativist world state: the initial states were stipulated to be exactly alike except for a mass doubling, and mass doubling does not affect the comparative mass relations at that time. Newton's laws were stipulated to be true at both worlds. Thus Newton's laws are not deterministic.

This is roughly how Baker presents the argument. But it is not quite satisfactorily formulated yet. Our discussion so far has been cast in absolutist terms. But to show that comparativism leads to indeterminism, we should speak in comparativist terms. This means, first, that the argument should be formulated to concern comparativist versions of the worlds Earth and Pandora. And it means, second, that when "Newton's laws" are stipulated to hold at the worlds, those laws must be Newton's laws *as the comparativist conceives of them*.

Now, in dealing with the second issue there is an obstacle: it is unclear what the comparativist conception of Newton's laws should be. Those laws as normally understood are mathematical equations relating absolute values of quantities, not as constraints on comparative relations; and as we'll see, there are difficult questions about how best to reformulate them in comparativist terms. So I suggest that Baker's argument be regarded as attacking any such conception, provided that the conception meets a certain constraint.

The resulting reformulated argument consists of three phases. Phase 1: begin by considering *absolutist* forms of Newton's laws, and a pair of *absolutist* worlds:

A-Earth: the absolutist laws of Newtonian gravitation hold; projectile escapes.

A-Pandora: same laws; same initial absolute positions and velocities; absolute masses are doubled; projectile returns.

By 'absolutist' worlds and laws I mean worlds and laws in which the metaphysics of quantity is the "flat-footed" one discussed earlier, according to which quantities like mass consist in the bearing of relations to real numbers. (On this view, Newton's laws are simply their textbook versions.) Next we move from these absolutist worlds and laws to comparativist correlates. We'll do this in two steps. The first step—phase 2 of the argument—moves from the absolutist worlds to a pair of comparativist *mosaics*—that is, totalities of fundamental matters of particular fact, across time and space, in comparativist terms. This is straightforward, since any absolutist mosaic induces a comparativist mosaic in an obvious way. (For example, if the comparativist admits a fundamental mass relation of $\succeq$, then $x \succeq y$ will hold in the induced mosaic if and only if the real number to which x bears the mass relation in the absolutist possible world is greater than or equal to the real number to which y bears the mass relation.) Call these mosaics the 'comparativist reductions' of the absolutist worlds. Next, in phase 3, we need to move from these comparativist mosaics to a corresponding pair of comparativist possible worlds whose laws are the comparativist's version of Newton's. (Given a Humean theory of laws the mosaic determines the laws, but the argument does not assume this.) The constraint mentioned above on the comparativist's conception of Newton's laws can now be stated:[28]

> On any adequate comparativist conception of Newton's Laws, there exist two worlds, call them C-Earth and C-Pandora, such that (i) the mosaics of C-Earth and C-Pandora are the comparativist reductions of A-Earth and A-Pandora, respectively, and (ii) Newton's laws under the comparativist conception in question are true in both C-Earth and C-Pandora.

Any comparativist Newtonian laws meeting this constraint fail to be deterministic. Any such laws are true in both C-Earth and C-Pandora. The states of C-Earth and C-Pandora are the same at the moment when the projectile is launched, since the mass-doubling does not affect the mass-relations at that time and nothing else is changed.[29] But C-Earth and C-Pandora have different world states at later times.

When the argument is formulated in this way, space for certain objections might seem to open up. According to the Humean view of laws of nature, a law is

---

[28] One can think of the constraint as saying that the operation of comparativist reduction transforms laws as well as mosaics: if you begin with an absolutist world, with certain laws L, and you apply the operation of comparativist reduction, then not only do you get a mosaic that is the comparativist version of the absolutist world's mosaic, but you also get laws that are comparativist versions of the absolutist world's laws—or better: you get a mosaic that is *compatible* with the existence and lawhood of comparativist versions of the absolutist laws.

[29] Baker does a nice job of circumventing certain obstacles to the argument raised by the "at-at" theory of motion.

a generalization in the "best" system, the system that achieves the best balance of strength and simplicity.[30] Since C-Earth and C-Pandora are simple possible worlds, a Humean about laws might claim that the comparativist Newtonian laws aren't *laws* in those worlds, because there isn't enough complexity for them to emerge as winners in the best-system competition. This isn't an objection to the argument as stated, since the argument required only that the comparativist laws be *true* at C-Earth and C-Pandora; but the Humean might argue for a weakened conception of determinism on which determinism is only falsified by such worlds if the laws in question are *laws* in those worlds. And even a nonHumean might object that the comparativist Newtonian laws aren't even true in C-Earth and C-Pandora, on the grounds that those laws make strong existence-assumptions of the sort discussed earlier, which assumptions are false in these simple worlds. However, I doubt that either objection is plausible: partly because it isn't clear that the worlds are simple in the way the objection requires, but more importantly because the indeterminism brought out by Baker's argument isn't limited to simple worlds. The simplicity of the example was convenient, but mass-doubling of a world state in almost *any* world, even a complex one like ours, will make a difference to the comparativist future given Newton's laws.

Baker's argument presents a formidable challenge to comparativism. To be sure, Newtonian gravitational theory has long been abandoned as a theory of our own world. But this is no comfort for the comparativist, since similar issues may well arise for successor theories usually thought to be deterministic. For that matter, even the determination of future chances in stochastic theories might be undermined by comparativism. Baker's argument shows that comparativism isn't "just metaphysics"; comparativist science is worrisomely different qua science.

This might seem to rest on the dubious methodological assumption that metaphysics must never meddle with science. The assumption is especially dubious in cases where the aspect of the science in question is based purely on metaphysical presuppositions that scientists make in ignorance of the alternatives, presuppositions that scientists are not particularly qualified to defend, and which can be dispensed with by alternate empirically adequate theories that are as attractive as the usual ones. But in the present case there are powerful reasons for even comparativists to recognize determinism, *empirical* reasons: we seem to have a large body of evidence suggesting that a system's behaviour depends on its initial conditions.[31]

Imagine we are committed comparativists and also ideal Newtonian physicists. Suppose further that we have tested Newton's laws over hundreds of years against as wide a range of data as possible, and have never found those laws to reliably fail. To be sure, we have recorded our observations and formulated the

---

[30] Lewis (1973, pp. 73–4; 1983, pp. 366–8; 1986b, pp. 121–4; 1994).

[31] Dasgupta (2020, section 4) argues that the empirical case for determinism is weightier than the a priori case.

laws numerically—in terms of SI units, suppose; and these numerical representations code up the comparative facts only indirectly, via representation functions. Nevertheless, we are in possession of a massive body of evidence in favour of Newton's laws, written down in terms of those representation functions. These laws make deterministic numerical predictions about the future, which code up predictions about the comparativist future, including predictions about whether fired projectiles will return to Earth or escape. We have justifiably high confidence in those predictions; we *know* when carefully constructed and fired projectiles will escape. In this scenario we would seem to have very powerful evidence for something like determinism; and even in less idealized scenarios we would seem to have reasonably strong evidence for that conclusion.

4.9 Lawhood relativized to representation functions

These thoughts point towards a way for a comparativist to secure a kind of determinism. But as we will see, it involves certain unattractive elements.

The first thing is for the comparativist to insist that the laws concern representation functions. Field would object, but as we saw in section 4.7, it is unclear whether his insistence on intrinsic laws can be sustained.

If the laws are to concern representation functions, it is natural to assume that, rather than concerning particular representation functions, they must quantify over representation functions, and in particular, say something about all representation functions. After all, it is arbitrary which function one picks to represent a given quantity.

But we cannot say simply that $\vec{F} = m\vec{a}$ (for example) holds for all representation functions, since the value 1 for the proportionality constant in this law is tied to particular units and thus to particular representation functions; and similarly for the constant G in the law of gravitation. It just isn't true that for all scales for measuring force, mass, and acceleration, the net force on any object is identical to the product of the mass of that body and its acceleration.

All right, one might think next: perhaps the laws say rather that for any representation functions, there exist *some* constants under which the familiar Newtonian equations hold.

To state this exactly, let's make some assumptions, for the sake of definiteness and simplicity, about the structure of space and time. These will be, roughly, Newton's own assumptions.[32] Space and time are substantival; points of space endure over time and have constant (better: not temporally relative) geometric relations; particles occupy points of space at times and stand in geometric relations derivatively, via the points of space they occupy. The comparativist treatment of space and time will thus make use of comparative predicates over points of time, comparative predicates over points of space, and an occupation predicate for a particle o occupying a spatial point p at a time t. Representation functions for

[32] All of what follows could, with a bit more complexity, be restated for Galilean space-time.

position at a time will be mappings $\vec{x}$ from pairs of particles and real numbers into $\mathbb{R}^3$; $\vec{x}(o, t)$ is a triple of real numbers representing the spatial position that particle o occupies at the time represented by the real number t. (Representation functions for net force will also be mappings $\vec{F}(o, t)$ from pairs of particles and reals into $\mathbb{R}^3$.) These representation functions will be constrained by the comparative and occupation predicates in obvious ways.[33] Letting $d(o, o', t)$ abbreviate the formula for the Euclidean distance between $\vec{x}(o, t)$ and $\vec{x}(o', t)$, that is, $\sqrt{(\vec{x}_1(o, t) - \vec{x}_1(o', t))^2 + (\vec{x}_2(o, t) - \vec{x}_2(o', t))^2 + (\vec{x}_3(o, t) - \vec{x}_3(o', t))^2}$, the law envisioned in the previous paragraph can now be stated as follows:

Representation-function Newtonian gravitation, existential form For any representation functions for net force, mass, and position, $\vec{F}$, m, and $\vec{x}$, *there exist* real numbers k and G such that for any particle o and real number t:

$$\vec{F}(o, t) = k\, m(o) \frac{d^2}{dt^2} \vec{x}(o, t) \qquad \text{and}$$

$$\vec{F}(o, t) = \sum_{o'} \frac{G\, m(o)\, m(o')}{d(o, o', t)^2} \hat{r}_{oo't} \qquad \text{(sum over all other particles o')}$$

(where $\hat{r}_{oo't}$ is the unit vector pointing from $\vec{x}(o, t)$ towards $\vec{x}(o', t)$).

I call this the 'existential form' because the constants k and G in the usual formulation of the laws have become existentially quantified variables. This feature, in fact, makes the existential form of the laws weaker in an important way than the absolutist's version.[34] In the absolutist's version, the gravitational constant G is physically significant—it is a matter of how strong the gravitational force is, relative to the amount of force needed to accelerate a particle of a given mass a certain amount. It affects what a planet's escape velocity is (recall its occurrence in the escape velocity formula). Given particular units, it has a specific, physically significant numerical value. (For example, its value is approximately 6.67×10^{-11} when force, mass, distance, and time are measured in Newtons, kilograms, metres, and seconds.) But in the existential form of the representation-function-theoretic law, G is just an existentially quantified variable. As a result, the law is indeterministic in a certain sense, since the values of the constants are needed to generate particular predictions about the future.

More fully, suppose we understand determinism in terms of representation functions, as follows:

---

[33] Here is one such constraint. Let $\vec{x}$ be a representation function for position (a function from particles and reals into $\mathbb{R}^3$); let τ be a representation function for time (a function from times onto $\mathbb{R}$); define $\mathrm{occ}(o, T)$ to be the point of space that particle o occupies at time T; define a function $\vec{X}$ from particles and times into $\mathbb{R}^3$ thus: $\vec{X}(o, T) = \vec{x}(o, \tau(T))$; let $\mathrm{dist}(\vec{x}_1, \vec{x}_2)$ be the Euclidean distance between $\vec{x}_1, \vec{x}_2 \in \mathbb{R}^3$. Then we can state a constraint on $\vec{x}$ by a four-place relation $\equiv$ of equidistance over points of space, as follows: for any particles o_1, o_2, o_3, o_4 and times T_1, T_2, T_3, T_4, $\mathrm{occ}(o_1, T_1)\,\mathrm{occ}(o_2, T_2) \equiv \mathrm{occ}(o_3, T_3)\,\mathrm{occ}(o_4, T_4)$ if and only if $\mathrm{dist}(\vec{X}(o_1, T_1), \vec{X}(o_2, T_2)) = \mathrm{dist}(\vec{X}(o_3, T_3), \vec{X}(o_4, T_4))$.

[34] Thanks to Verónica Gómez and Ezra Rubenstein here.

Some representation-function laws are deterministic if and only if for any world at which they are true and any representation functions $f_1, \ldots$ at that world, a complete description of that world at any time using $f_1, \ldots$ plus those laws implies any complete description using $f_1, \ldots$ of that world at any other time.

('Implies' here means mathematical rather than modal implication: the laws plus the first world-state description plus the laws of mathematics logically imply the second world-state description.) The existential-form laws are indeterministic in this sense. Choose a world in which they are true, and pick some representation functions to measure force, mass, distance, and time, and describe an initial state of the universe numerically, in terms of those functions, with description D_0. For any other time, the law needn't imply the numerical description D of the world at that later time, in terms of those same representation functions. D and D_0 are numerical descriptions of world-states, and whether one numerical description of a world state plus the numerical versions of Newton's laws yields another numerical description of a world state depends on the values of the constants in those laws, the gravitational constant, and the constant in the law of motion. D_0 plus Newton's laws with one pair of constants will yield one later world state; D_0 plus Newton's laws with another pair of constants will yield a different world state (unless the chosen world happens to be extraordinarily simple, in which case choose another that isn't). Thus D_0 plus the statement that Newton's laws hold with *some pair of constants or other* (which is what the above law says) is consistent with each such later state.

We have hit a roadblock. But here is a different way to think about representation-function-theoretic laws, which I call the *relativizing approach*. According to this approach, instead of trying to state laws simpliciter, we rather state laws *relative to a given choice of representation functions*. (We'll return to the significance of this shift.) According to the relativizer, for any particular representation functions for net force, mass, and position, $\vec{F}$, m, and $\vec{x}$, there is some law, relative to these choices, of this form:

Representation-function Newtonian gravitation, relative to $\vec{F}$, m, and $\vec{x}$

For any particle o and real number t:

$$\vec{F}(o,t) = k\,m(o)\frac{d^2}{dt^2}\vec{x}(o,t) \qquad \text{and}$$

$$\vec{F}(o,t) = \sum_{o'} \frac{G\,m(o)m(o')}{d(o,o',t)^2}\,\hat{r}_{oo't} \qquad \text{(sum over all other particles } o'\text{)}$$

(where k and G are *particular* real numbers).

That is, relative to any chosen representation functions, there will be some *particular* pair of constants that appear in laws of gravitation and motion for those chosen functions.[35]

Given the relativizing approach to lawhood, determinism could still hold in a Newtonian world, despite Baker's argument, if determinism is understood as follows:

Relativization determinism A world is deterministic if and only if for any representation functions $f_1, \ldots$ at that world, a complete description of that world at any time using $f_1, \ldots$ plus the world's laws relative to those representation functions implies any complete description using $f_1, \ldots$ of that world at any other time.

(Note that it is now worlds rather than laws that are said to be deterministic.) For suppose Representation-function Newtonian gravitation holds at C-Earth, consider any representation functions for force, mass, position, and time, and consider any description of that world at the initial time (with the projectile at the Earth's surface) in terms of those functions. There will be laws relative to the chosen representation functions for force, mass, and position, with particular constants, one for the law of dynamics and the other for the law of gravitation. These laws are identical in form to the usual absolutist sort. So we can derive a description, in terms of these same representation functions, of that world at any later time, from the initial description, by evolving the earlier numerical description using the law in the usual way. Determinism as understood above is thus true.

Instead of entirely giving up on laws-simpliciter it might be better for the relativizer to say instead that the existential form of the law is indeed a law simpliciter, but that it isn't the full nomic story; the full nomic story includes also the relativized laws. It is important to appreciate that there is great pressure on a comparativist to admit that the statements I am calling relativized laws are indeed laws in some sense. For they encode information—for instance, information about which projectiles will escape the gravitational fields of planets—which is clearly universally applicable, general, projectible into the future, epistemically available to and of interest to scientists in a comparativist world, and so on. And this information is not encoded in the existential-form laws.[36] As emphasized at the end of the previous section, there is powerful evidence in a Newtonian world for the Newtonian laws in their usual form, in SI units for example. This evidence is for something stronger than the indeterministic existential-form laws-simpliciter; it is evidence for the Newtonian laws relative to the representation functions for

---

[35] Martens (2017b, section 5) considers a related idea.

[36] Similarly, relativized laws are stronger than Dasgupta's (2020) "minimalist" laws, which constrain two objects at a time, saying for example that if one thing is twice as massive as another and is acted on by the same force, it will experience half the acceleration of the other.

SI units (and also is evidence for Newtonian laws relative to other representation functions).

4.9.1 *Epistemic objection*

If a sort of determinism can hold given comparativism, what, then, breaks down in Baker's argument? Well, his argument wasn't directed at determinism in the sense defined above (Relativization determinism), so nothing needs to break down. But what, intuitively, is going on? As we saw, we can derive numerical descriptions of later times from earlier ones, given the relativized laws. But how can that be? The later descriptions determine whether the projectile escapes, and the two worlds are exactly alike, in a comparativist sense, at the earlier time.

A representation function as a whole encodes information about objects at all times. In order for a function, m, to count as a mass function, for example, it must be that for any objects x and y, regardless of their locations in time, $x \succeq y$ if and only if $m(x) \geq m(y)$. Such a function encodes facts about comparative relations between objects on all time slices of the world, and also comparative relations between objects in different time slices. So when it comes to making predictions about what a certain time slice will lead to, a law that makes reference to such functions has more to work with, so to speak, than merely facts about the intrinsic features of that time slice. It also has access, in the numerical description of that time slice, to facts about that time slice in relation to other time slices.

This admittedly represents a significant departure from our ordinary conception of laws. We ordinarily think of Newton's second law as being local in a certain way: in order to tell us how an object will move, it need "look" only at that object's immediate spatiotemporal vicinity. Laws about representation functions aren't local in this way.[37] (Nor were Field's intrinsic laws, as we saw in section 4.7; it is hard to see how laws on any comparativist conception could avoid such nonlocality.) But I'm not sure I agree with Baker when he says that 'A notion of determinism in which the "initial conditions" include information about the future would thus appear remarkably ill-suited to the epistemic role determinism normally plays in science.'[38]

The suggestion, perhaps, is that we could never be justified in believing a representation-function-theoretic law. To do so we would need to observe cases in which it correctly applies, but such cases involve numeric descriptions of times, and numeric descriptions are "future-infected": we could never know them without knowing what the future will be like, and we cannot know what the future will be like if we don't already know the law.

Thus understood, the objection recalls an old dialectic. According to the Humean about laws, what makes a generalization a law is its inclusion in the best

---

[37] Despite having the potential to be local in another sense; see note 19.

[38] Baker (2020, p. 88). Baker says this about a different proposal for comparativist laws which makes reference to representation functions.

summary of the nonnomic facts across all of time and space. Thus understood, laws are facts that involve the future, much in the way that the "initial conditions" on the representation function approach involve the future. Some people—such as D. M. Armstrong (1983, chapter 4, section 5)—object to Humeanism on that basis, saying that it's incompatible with the idea that we can be justified in believing the law on the basis of observation. Humeans reply by conceding that they must rely on some sort of nondeductive epistemic magic to move from past observations to Humean laws (and thus, in effect, from past observations to future observations), but ask why that is any worse than the nondeductive magic that nonHumeans like Armstrong require to move from past observations to nonHumean laws.

The dialectical situation with the representation-function approach is similar. Our usual, numerical, representation of the present state of the world is in a sense future-infected, since the numerical representations are partly constituted by relations to future objects. But our epistemic access to that numerical representation ultimately derives from something that is not future-infected and is epistemically less problematic: comparative relations amongst present and past things. Once we have set numerical scales for mass, distance, and so forth, by deciding which objects are to be assigned the value 1 for those quantities, our assignment of numerical values to other things is ultimately based on comparisons: we compare masses with sets of scales, distances with rulers, and so forth. I say 'ultimately' because much of our numerical representation of the current state of the world is based on our knowledge of the laws, as when we estimate the mass of the sun by appeal to the laws of gravitation. Still, the comparativist is in a similar situation to the Humean here: she must appeal to the same sorts of nondeductive forms of inference (appeals to simplicity, etc.) that the Humean requires, for moving from our knowledge of present and past comparative relations to a battery of conclusions about the numerical representation of past and present states of the world and about the laws.[39] If we were in a very simple world like C-Earth, perhaps we wouldn't then know much about the future (for example, whether the projectile would return), but in realistic cases, in worlds that are complex like our own, we have enough information to know about the future. In such worlds we have had experience with a large number of systems, which confirms, given any chosen representation functions, a numerical law that will decide questions about the future in one way or another. In particular, in a complex enough world, relative to any chosen representation functions, we'll have enough information to determine the values of the gravitational and dynamical constants, relative to those functions.

Baker's epistemic-role complaint might instead be that representation-function laws cannot explain. Explanation, he might insist—nonprobabilistic explanation,

---

[39] In a defence of relationalism, Dasgupta (2020) claims that our evidence does not support determinism (in a non-relativization sense) since the conclusions we draw about the future are not based on intrinsic features of a single past or present moment, but rather are based on features from multiple times. In part this is in the spirit of what I am saying here, though Dasgupta does not defend relativization or representation-function laws.

anyway—requires nonrelativized laws and determinism in the original sense. But this complaint would seem to be dialectically ineffective. The relativization approach admittedly fails to deliver a certain ideal sort of explanation. But if representation-function laws are supported by the evidence and predict the future, why not say that they also explain in some good sense?

4.9.2 *Simplicity objection*

Baker objects to the idea that laws quantify over representation functions, saying that laws must be statements about fundamental properties:

It is a familiar platitude that, while fundamentality may be a brute concept with no definition, the fundamental properties and relations 'are the properties and relations that occur in the fundamental laws of physics' (Arntzenius, 2012, p. 41). On Lewis's popular Humean account of laws, for example, the fundamental laws are regularities in the instantiation of fundamental properties (Lewis, 1983, p. 368). And altering this feature of Lewis's system would rob it of much of its interest. Our best theories of physics have a particular mission: to describe the universe at its most fundamental level. Insofar as they fail to do so, either through inaccuracy or by failing to describe reality in fundamental terms, we should take that as a sign that the true fundamental laws have not yet been discovered.

(Baker, 2020, p. 87)

Now, what Baker is objecting to here is not the relativizing approach, but rather the view that there are laws simpliciter that quantify over all choices of representation functions. And against that sort of view, the objection fails. For consider the existential form of Newton's laws, which quantify over all choices of representation functions (rather than mentioning any one in particular). That *is* a simple statement about fundamental properties and relations. When you unpack the notions of a mass-function, a position function, and so forth, the result makes reference only to comparative relations like $\succeq$ and C, which are fundamental according to the comparativist, plus mathematical concepts, which themselves may well be fundamental. The mention of representation functions in laws thus does not on its own contradict the assumption that laws can mention only fundamental properties and relations, nor does it require a departure from the laws-centric epistemology of fundamentality.

Baker's objection is more appropriate as directed against the relativizer, since relativized laws concern particular representation functions. Such a function is given in extension, and does not have a simple definition in fundamental terms. However, there is a sense in which the argument is unconvincing even here. The relativizer does not claim that the equations governing particular representation functions are laws simpliciter; they are only laws relative to their representation functions. Insofar as anything in the relativizer's outlook counts as a law simpliciter (setting aside the existential-form laws), it is the entire set of relativized laws; and there is a sense in which this set constitutes a simple constraint on the fundamental comparative relations. Let the quantities in the world in question be $Q_1, \ldots, Q_n$. For each quantity there is a notion of a representation function; call a *choice* of representation functions an n-tuple of representation functions, one for each quantity: that is, some $\langle f_1, \ldots, f_n \rangle$ where each f_i is a representation function for Q_i. For each choice there are some relativized laws; call the set of all relativized laws for all choices $\mathscr{L}$. Now: each member of $\mathscr{L}$ is a constraint on the fundamental comparative relations (since facts about representation functions constrain the comparative relations in terms of which those functions are defined). Moreover, each member of $\mathscr{L}$ is a *simple* statement about its corresponding choice. Furthermore, the entire set of choices has a simple definition in terms of the fundamental comparative relations (recall the simple definition of a mass function in terms of $\succeq$ and C). And finally, the members of $\mathscr{L}$ stand in simple mathematical relationships to one another. Consider any two choices, $R = \langle f_1, \ldots, f_n \rangle$ and $R' = \langle f_1', \ldots, f_n' \rangle$. Given Uniqueness theorems, each pair of functions f_i and f_i' will be related by "transformation constants". For example, if f_i and f_i' are mass functions then for some real number k, for all objects x, $f_i'(x) = k f_i(x)$; the multiplicative constant k is the transformation constant in this case.[40] In terms of these transformation constants, there will be a simple mathematical rule for transforming the laws relative to R into the laws relative to R'. The laws always have the same form, and differ only in terms of their constants (such as the gravitational constant G); and the constants' values for the laws relative to R' are determined by a simple function from those constants' values in the laws relative to R and the transformation constants.

Thus there remains a sense in which the relativizer's laws are simple statements about the fundamental properties, and a corresponding sense in which this approach complies with the laws-centric epistemology.

To further defend relativization from the objection, consider the matter from the perspective of the Humean view of laws. From that perspective the assumption that laws are simple statements about fundamental properties and relations is no mere dogma. It rather plays an essential role in avoiding the trivialization of lawhood. As Lewis pointed out, without any constraints on the lawmaking language we could choose a predicate F that is true only of objects in the actual world and formulate an extremely simple and strong system with just one axiom, $\forall x F x$. This would count as the best system, which would make all true generalizations be laws. Lewis avoided this trivializing result by requiring that all primitive predicates in the lawmaking language express fundamental properties and relations.

But in fact the Humean account of laws can be reworked under the assumption of relativization. As we will see, trivialization is avoided in essentially the same

---

[40] The nature and number of transformation constants depends on the kind of quantity at issue. For a (real-valued) "ratio scale" like mass, the transformation constant is a single real number. For an interval scale—a scale for which it is ratios between intervals that are physically significant—for any two representation functions f and g there will be two transformation constants k_1 and k_2 such that for all x, $f(x) = k_1 g(x) + k_2$. Other quantities might differ in the nature of the scale, in the number of argument places, and in the kinds of mathematical entities that representation functions map to, and thus will have correspondingly different transformation constants.

way as it was for Lewis, despite the fact that individual relativized laws are not simple statements about the fundamental properties and relations. The idea for the reworking will be this: instead of looking at the outset for simple and strong statements in terms of the fundamental properties and relations, we will look for simple statements only after picking representation functions.

Let us ignore probabilistic laws for simplicity. I assume that each fundamental quantity Q has a certain number of argument places n, and is associated with some fundamental comparative relations $R_1, \ldots, R_m$, which determine a class of representation functions for Q in any possible world w.[41] These representation functions are mappings from n-tuples of objects from w to mathematical entities that are constrained by the relations $R_1, \ldots, R_m$. Example: mass is a fundamental quantity with one argument place, associated with fundamental relations $\succeq$ and C; and a representation function for mass in any world w is a mapping f from 1-tuples (i.e. objects) in w to real numbers such that for any x, y, z in w, $f(x) \geq f(y)$ if and only if $x \succeq y$ and $f(x) + f(y) = f(z)$ if and only if $Cxyz$.

Where N is any numerical statement of the sort that the relativizer calls a relativized law, let its *matrix* be the open sentence in which all names for representation functions are replaced by variables. Although N is not a simple statement about the fundamental properties and relations (since it is about particular representation functions), its matrix may be simple.[42]

Here, then, is the Humean's definition of the relativized laws at an arbitrary possible world, w. Let $f_1, \ldots, f_n$ be any representation functions for the fundamental quantities at w, and consider various axiomatic systems in a language with no nonlogical vocabulary other than names for the representation functions $f_1, \ldots, f_n$, predicates for nonquantitative perfectly natural properties and relations, and mathematical vocabulary. The laws relative to $f_1, \ldots, f_n$ are the generalizations in the system that best balances strength with the simplicity of its axioms' matrices.[43]

On this view there is no trivialization: trivialization is avoided by the step where we only consider representation functions defined (using measurement theory) from the fundamental properties and relations. So insofar as 'laws concern only fundamental properties' derives from the desire to avoid trivialization of Humean laws, that slogan should not tell against the relativizer's approach at

---

[41] Notice the appeal to a notion of representation-function-for-Q, for arbitrary Q. It is clear in practice what the representation functions are in the particular cases with which we are familiar (e.g. mass), but there is a question of the legitimacy of the general notion (and perhaps room for a bit of conventionality to creep in).

[42] Assuming that the presence of a small number of names for particular real numbers—the constants in the law in question—does not make the statement horribly complex. (It might make it somewhat complex; we can all admit that reducing the number of fundamental constants is a goal of ideal physics.)

[43] Is there any reason to evaluate strength by reference to matrices (somehow quantifying over various possible assignments to those variables), rather than directly evaluating the strength of the theory itself, which is about particular representation functions?

all. Put another way: at any world, the Humean can define, by appeal only to fundamental properties and relations, the class of relativized laws at that world.

4.9.3 *Nomic quotienting*

The epistemic and simplicity arguments are unconvincing. Is the relativizing approach, then, an appealing one? It seems to me that it is quite *un*appealing, but in a certain subtle and distinctive way. According to the approach, the laws cannot be stated without first making an arbitrary choice of representation functions. To bring out how odd this is, imagine pressing the relativizer to say what it is about the world that selects what the laws are relative to a given chosen set of representation functions. She will need to say that there is no simple way to answer the question. She might say that 'the world is such as to make *these* the laws under these choices, *those* the laws under those choices', and so forth. But given how mathematically parallel all the laws are to one another, this is an intuitively unsatisfying position. There ought to be, one is inclined to think, some simple explanation of why the laws keep coming out in the same mathematical form, and why the particular constants that "pop into place" do so, whenever a set of representation functions are chosen.

In a Newtonian gravitational world, if you pick particular representation functions, then the laws will involve some particular values for the two constants in the law of motion and the law of gravitation. Under other choices of representation functions, the laws will have the same form but will have different constants. But there won't be some "uber law" that accounts for these different relativized laws, since the uber law would need to make reference to absolutist structure that isn't there. The best that we can say about the world's nomic structure is that it is such as to yield these laws relative to these choices, and those laws relative to those choices. We can say that the world has the potential to yield any of the particular relativized laws given their choices, but we cannot articulate this potential in other terms.

The relativizer is analogous to a character that I call the 'quotienter', whom I have mentioned before and will introduce more fully in Chapter 5. The quotienter accepts that there are multiple, equally good ways to describe the world, but denies the need to articulate what it is about the world that makes those different descriptions equally good. For example, if Leibniz had been a quotienter, he would have said:

I am happy to accept the existence of substantival space, just as my friend Newton does. However, I disagree with his assumption that there is a single best description of bodies' locations in that space. Rather, there are many equivalent, equally good descriptions. We can describe the totality of bodies as being located *here* in space, or we could equivalently describe them as being located *there*, or at any other total location that preserves the inter-body distances. But there is no way to say what it is about the world that makes all these descriptions equivalent. All we can say is: the world is such as to be well represented either this way or that.

(Leibniz himself was of course not a quotienter. For he *did* say what makes all these descriptions equivalent: the spatial facts for him are ultimately not about the occupation by bodies of points of space, but rather about the holding of spatial relations between bodies.) As we will see, there are many other important instances of this attitude, and it is an important meta-stance underlying many foundational disputes in metaphysics and philosophy of science, including many that we have been discussing in this book.

I see the relativizer as engaging in a kind of "nomic quotienting". The relativizer perhaps concedes that the world's *mosaic* can be given a nonquotienting description (and thus need not be a quotienter simpliciter[44]), but says that the *laws* cannot be given a nonquotienting description. There is simply no way to say what is nomically fixed in any uniquely best way. Rather, we choose a vocabulary, and then the nomic facts magically snap into place, just as, for the quotienter simpliciter, we choose some arbitrary parameters for describing the world—the locations in space of three non-colinear point particles, say, in the case of quotienting-Leibniz—and then all the other nonnomic facts under that choice—the locations of all other bodies in space—magically snap into place.

Here is one further illustration of the idea of nomic quotienting, from mathematics rather than physics. In set theory the axiom of choice is often stated in terms of functions, for instance:

> Any set X of (perhaps overlapping) nonempty sets has a "choice function", a function that assigns to any member of X an element of that member.

But functions are not primitive entities of set theory. Rather, they are normally defined as certain sets of ordered pairs, and ordered pairs are in turn defined as certain unordered sets. But there are many equally good ways to define ordered sets in terms of unordered sets. Thus there is an element of arbitrariness in the above statement of the axiom of choice: any articulation of this law of set theory in terms of primitive vocabulary (that is, in terms of the predicate '$\in$' of set membership) must be based on an arbitrary choice. There is something intuitively problematic here. Pick any suitable method for coding up ordered pairs and you get a corresponding version of this axiom of choice; there ought to be, one is inclined to think, some explanation of why these laws—all sharing a certain form—keep popping into place.

But in fact such an explanation *can* be given in this case, and so nomic quotienting in set theory can be avoided.[45] Amongst the many equivalent statements

---

[44] Although notice if that if the relativizer is an antireductionist about laws then quotienting simpliciter will be required after all, since the fundamental description of the world will then include a specification of the laws.

[45] In first-order set theory, that is. Arbitrariness returns if the axiom of replacement is formulated using plural quantifiers over ordered pairs, as in Boolos (1984, p. 448). But it is banished again if that axiom is formulated in full (nonmonadic) second-order logic.

of the axiom of choice, some do not appeal to functions or ordered pairs. There is this one, for instance:

> For any set X of nonoverlapping nonempty sets, there is a set containing exactly one member from each member of X.

The multiplicity of equally good versions of the choice-function formulation of the axiom of choice (corresponding to different methods for defining ordered pairs) may be given a simple explanation, since each is derivable (once a method for defining pairs is chosen) from this second version of the axiom of choice, which has no analogous arbitrariness.

4.9.4 *Inter-world obedience of laws*

There is a further worrisome feature about the relativizing approach: it seems to imply that the laws are not physically meaningful when applied to counterfactual scenarios.

It is essential to the conceptual role of laws that they be applicable in counterfactual scenarios. For instance, in our ordinary evaluations of counterfactuals, we assume that the actual laws would have continued to hold under most counterfactual suppositions. Thus it must make sense to speak of the laws from the actual world holding in counterfactual scenarios. More generally, "inter-world obedience" of laws needs to make sense: for any possible worlds w and w', we must be able to speak, in a physically meaningful way, of whether the laws of w hold at w'.

But on the face it, we cannot do this for relativized laws. Laws-relative-to-representation-functions-$f_1,\ldots$ in world w are statements about the particular representation functions $f_1,\ldots$, and are not physically meaningful when applied to other possible worlds w'. For a representation function is given in extension, and has no physical significance in other worlds. It's only in w that $f_1,\ldots$ count as representation functions, and thus it's only in w that those functions are guaranteed to be correlated with the comparative relations.

(A function that counts as a mass function, say, in one possible world, w, needn't count as a mass function in another world v. Suppose that in w, $Cxxy$ for some objects x and y. Then any mass function for w will assign to y a number that is twice what it assigns to x. Let f be such a function; suppose that $f(x)=1$ and $f(y)=2$. But now let v be some world in which $Cxxy$ is not true. f therefore does not count as a mass function for v, since $f(x)+f(x)=f(y)$ even though $Cxxy$ is not true in v. Thus even though the statement that $f(y)$ is twice $f(x)$ is true in v—since functions do not, I assume, assign different values to their arguments relative to different worlds—it is not a physically meaningful statement about v.)

The problem is not due solely to the presence of representation functions in laws. Representation-functional laws in the existential form, for example, do not have the problem: since they quantify over representation functions rather than mentioning particular ones, they can meaningfully be applied to multiple possible worlds: they make a statement, for any world, about the nature of whatever functions count as representation functions *in that world*.

Ultimately, though, I don't think the problem is fatal: if we could learn to live with nomic quotienting itself, we could limp along here too. I'll mention two ways of doing so.

The first is counterpart-theoretic.[46] Begin with a world, w, with laws L relative to representation functions $f_1,\ldots$ for w. For certain other worlds, v, there will be a natural and unique way for choosing representation functions $g_1,\ldots$ that can be viewed as the counterparts of $f_1,\ldots$ in v. For example, if v has a large portion (an initial segment, say) that is an intrinsic duplicate of some portion of w, then we can construct the counterpart functions $g_1,\ldots$ by first having them assign the same values as $f_1,\ldots$ within the duplicated portion, and then extending them to the rest of v as dictated by the comparative quantitative relations amongst the objects in v. Corresponding to these counterpart representation functions, we can formulate a counterpart to L: the statement that $g_1,\ldots$ stand in the numerical relationship that L attributes to $f_1,\ldots$ (the relationship whose definition in L includes the constants L mentions). A world v that has such counterpart representation functions can be said to obey L by proxy, if and only if the counterpart to L is true in v.

Some worlds will not have unique counterpart representation functions. Assume the actual world is Newtonian; choose representation functions that correspond to the standard units, so that the laws for those functions take the usual form, with the usual constants (1 for the constant in the dynamical law, and 6.67×10^{-11} for G in the law of gravitation); and consider a very simple world, containing a single particle moving alone in space. This world doesn't contain a sufficiently large and rich duplicate of any actual region that would determine unique counterpart representation functions. In such a case, the only statement that we might regard as a counterpart to w's relativized law L would seem to be the weaker—and nondeterministic—existential-form statement, the statement that for any representation functions, Newton's laws hold for *some* pair of constants. Such a world, v, may again be said to obey L by proxy if and only if this counterpart to L is true in v; but now the counterpart statement is weaker. A similar problem arises even in complex worlds if those worlds are too dissimilar from our world.

Thus we have a notion of w's relativized laws being obeyed in other worlds, although the obedience is more lax if the other worlds are simple or too dissimilar from w.

A second way of dealing with the problem can be based on an idea of Dasgupta's, a distinction between "strict" and "loose" possibility.[47]

---

[46] Dasgupta (2013, section 3) defends a similar approach; see also Lewis (1986a, pp. 70–1).

[47] Dasgupta (2020). Dasgupta develops the ideas in defence of a different comparativist response to the problem of indeterminism. Though there are important differences between his response and relativization, there are important similarities as well. My development of loose possibility is not exactly the same as Dasgupta's.

The strict possibilities are the various configurations of fundamental reality—for the comparativist, distributions of the fundamental comparative relations over all the individuals.

The loose possibilities, on the other hand, are in a sense fictional. Pretend that there exist absolute magnitudes of the sort that the mixed absolutist accepts—fundamental determinate properties and relations of having exactly *this* mass, of being separated by exactly *this* distance, and so forth—and that each number assigned by a representation function corresponds to one of these magnitudes. For instance, suppose a certain mass function, m, assigns the number 1 to me and the number 2 to Shaquille O'Neal. (It follows that the $\succeq$ relation holds between O'Neal and me, and that the concatenation relation C holds between me, me, and O'Neal.) Under the pretence, there are a pair of absolute mass properties, p_1 and p_2, corresponding to the numbers 1 and 2, which are possessed by me and O'Neal, and which are such that $*(p_1, p_1, p_2)$. Now, if these absolute magnitudes really existed, there would also exist possible worlds (strict possible worlds, in fact) in which objects possess different absolute magnitudes from those that they actually possess. There would be one, for instance, in which I have the property p_2 and O'Neal has p_1. So corresponding to the fictitious absolute magnitudes, there is a space of fictitious possible worlds. We can represent these fictitious possible worlds by functions that assign numbers to individuals representing the fictitious absolute magnitudes possessed by the individuals in those worlds. For instance, such a function corresponding to the fictitious world just imagined would, in the case of mass, assign the number 2 to me and the number 1 to O'Neal. These functions are the loose possible worlds.

More carefully, relative to any choice of representation functions, $f_1, \ldots$ for all the fundamental quantities, a loose possible world may be defined as a function that assigns a number to any pair $\langle f_i, \vec{o} \rangle$, where f_i has n places and $\vec{o}$ is an n-tuple of individuals. Under the pretence, a loose world w can be thought of as a possible distribution of the fictitious absolute magnitudes over all individuals; $w(f_i, \vec{o}) = r$ represents the objects $\vec{o}$ having the absolute magnitude that f_i's assignment of r represents. For instance, the loose possible world imagined at the end of the previous paragraph would be a function w such that $w(m, \text{Sider}) = 2$ and $w(m, \text{O'Neal}) = 1$.

We can now introduce the idea of a numerical sentence about quantities being true at a loose world. Return to the mass function, m, such that $m(\text{Sider}) = 1$ and $m(\text{O'Neal}) = 2$. Mass functions, recall, are constrained by the *actual* distribution of the comparative relations; the values they assign have no physical significance in possible worlds in which the comparative relations are differently distributed. Still, we may count the sentence '$m(\text{Sider}) = 2$ and $m(\text{O'Neal}) = 1$' as being true at the loose possible world w at the end of the previous paragraph, because $w(m, \text{Sider}) = 2$ and $w(m, \text{O'Neal}) = 1$. Once this idea is extended to sentences of arbitrary complexity in the obvious way, we can speak of the truth of any sentence about representation functions in any loose possible world. And thus we can speak of representation-function laws being obeyed in loose possible worlds.

On this second approach, then, it is only when we speak of possibility in the loose sense that we may speak of relativized laws being obeyed in other possibilities.

4.10 Absolutism and laws

It is natural to think that if relativization is the best a comparativist can do, then so much the worse for comparativism. But this presupposes that there is a better absolutist alternative.

We were led to relativization by Baker's argument; and it might seem obvious that this argument is no threat to absolutism, since the absolutist's initial state of the world changes if all masses are doubled. But it isn't obvious that the laws as the absolutist conceives of them will be sensitive to this change in the world's state. Whether they are depends on what exactly those laws look like. There certainly are *some* conceptions of absolutist laws on which they would indeed be sensitive to the change. For instance, an absolutist might, in a Newtonian world, embrace infinitely many dynamical laws, each one specifying what the acceleration of a particle would be in one particular, completely specific situation; and similarly for "the" law of gravitation. But this is an unattractive conception of the laws, for the reasons given in section 4.4. As we will see, there are more attractive conceptions available that are sensitive to the doubling, but they have their warts; and other conceptions that lack the warts are insensitive to the doubling.

We cannot be sure that absolutism provides a better response to Baker, or, more generally, that it provides a more attractive conception of the laws, until we have investigated in detail what absolutist conceptions of the laws are available. And in fact, the difficult questions about quantitative laws that we have been asking of comparativists—whether they are intrinsic or representation-function-theoretic, and if the latter whether the relativizing approach is correct—arise for absolutists as well. Let us look at these questions as they confront absolutism, taking Mundy's mixed absolutism (section 4.7.3) as our working form.

One conception of laws open to a mixed absolutist is relativization. Relativized laws for the absolutist will be mostly like the comparativist relativized laws discussed in section 4.9, except that the representation functions to which the laws are relativized will now assign numerical values to the absolutist's determinate properties and relations, rather than to concrete individuals. Baker's argument for indeterminism won't work given this conception of laws (and given the corresponding understanding of determinism), for the same reason as before: given any representation functions, one can simply calculate the later world state from the earlier world state using textbook physics. But is this solution to Baker's problem any more attractive than the analogous comparativist solution based on comparativism?

Basing relativization on mixed absolutism rather than comparativism does avoid Baker's objection that initial conditions understood in terms of representation functions are future-infected (section 4.9.1). For an absolutist's description of a time-slice using representation functions is purely a function of the determinate absolute properties and relations instantiated on that time slice.

To see this in detail, consider how we attribute a numerical value to the spatial distance between objects o_1 and o_2 at some time, T. The fundamental absolute spatial relations for the mixed absolutist, let us suppose, are binary distance relations between points of space. (Each such relation is a nonnumeric relation of being exactly a certain distance apart.) These relations stand in higher-order structuring relations, analogous to the relations $\geq$ and $*$ over determinate masses from section 4.7.3; and a representation function for distance will then be a function $\mathscr{D}$ that assigns a real number to each such binary distance relation, in a way constrained by the higher-order structuring relations. Notice that this definition of a representation function makes no reference to the concrete world at all, and thus no reference to facts about times other than T; it makes reference only to facts about the space of binary distance relations. We then can say that the numerical value of the distance between o_1 and o_2 at T, relative to $\mathscr{D}$, is $\mathscr{D}(R(o_1, o_2, T))$, where $R(o_1, o_2, T)$ is the binary distance relation that holds between the points of space that o_1 and o_2 occupy at T. Note again that we still have made no reference to any time other than T.

Basing relativization on mixed absolutism also avoids the problem of interworld obedience of laws (section 4.9.4). For the relativized laws are now relativized to representation functions for properties and relations, which have physical significance across different possible worlds since they are defined without reference to concrete goings-on.

To see this, consider, for example, Newton's laws. Representation functions now assign mathematical entities to mass properties, binary relations of spatial distance, binary relations of temporal distance, and force properties or relations; and they do so solely as a function of the higher-order structure of those properties and relations. Now, these "higher-order" representation functions, together with certain first-order facts—namely, the facts of instantiation of the absolute properties and relations by particles, and the facts of particles occupying points of space at times—induce unique representation functions of the old sort, from concrete objects to mathematical entities. For instance, where $\mathscr{M}$ is a higher-order representation function for mass, there corresponds the "first-order" representation function $m(o)$ which assigns to any particle o the real number $\mathscr{M}(p)$, where p is the determinate mass property instantiated by o. Newton's laws, relative to any choice c of higher-order representation functions, can then be stated as follows:

The functions $\vec{x}(o, t)$, $\vec{F}(o, t)$ and $m(o)$ induced by c are such that for any particle o and real number t:

$$\vec{F}(o,t) = km(o)\frac{d^2}{dt^2}\vec{x}(o,t) \qquad \text{and}$$

$$\vec{F}(o,t) = \sum_{o'} \frac{Gm(o)m(o')}{d(o,o',t)^2}\hat{r}_{oo't} \qquad \text{(sum over all other particles } o'\text{).}$$

(Again, k and G are *particular* real numbers. This is not a law simpliciter; it is a law relative to choice c.)

Now, suppose such a statement to be, in the actual world, a law relative to c. This statement is also physically meaningful when applied to other worlds. The problem of inter-world obedience from section 4.9.4 was that the comparativist's relativized laws concerned particular functions, $\vec{x}(o,t)$, $\vec{F}(o,t)$ and $m(o)$, which were defined by reference to the actual world and lacked physical significance in other worlds. What functions does the mixed absolutist's relativized law concern? First, it names particular higher-order representation functions—the ones schematically indicated by 'choice c'. But these are physically significant in other worlds since, as noted, their definitions make no reference to concrete goings-on, only to the structure of the space of absolute properties and relations.[48] Second, it speaks of first-order functions. But these functions are not named; they are, rather, picked out by description, as the functions, whatever they are, that are induced by c; and the "inducing" involves not only the facts about the functions c, but also certain first-order facts; and the first-order facts that are relevant, when speaking of another world, w, will be first-order facts *about* w. Thus when evaluating the truth of the law with respect to world w, the definite description 'the functions $\vec{x}(o,t)$, $\vec{F}(o,t)$ and $m(o)$ induced by c' will pick out functions that reflect the positions, forces, and masses of particles *in* w. The situation is different with the comparativist's relativized law from section 4.9, for there, '$\vec{x}(o,t)$', '$\vec{F}(o,t)$', and '$m(o)$' were *names* of particular functions, and thus, when the law is evaluated with respect to another possible world, the functions denoted by those names reflect the positions, forces, and masses of particles in the *actual* world, not that other world.

Basing relativization on mixed absolutism, then, avoids the epistemic and inter-world-obedience concerns. But it doesn't avoid what to my mind was the central concern, which is that relativization is an objectionable sort of "nomic quotienting".

What non-relativizing conceptions of the laws are available to the mixed absolutist? There are, first, representation-function laws in the "existential form". But these are indeterministic in the case of Newtonian gravitational theory, despite the fact that the representation functions are now constrained by the absolutist's structuring relations over absolute magnitudes rather than the comparativist's relations over individuals. For the reasoning given above that they are indeterministic did not turn on the representation functions being defined in the comparativist rather

---

[48] They are not physically significant in worlds where the properties and relations are differently structured. But it is natural to deny that they are physically significant in such worlds.

than mixed absolutist way; it turned only on the constants being existentially quantified.

There is a conception of representation-function laws available to the mixed absolutist that is stronger than the existential form, but does not involve relativization. On this conception the laws quantify universally over all choices of representation functions; but particular, arbitrarily chosen, determinate properties are used to pick out the values of the constants in the laws, relative to any choice. Let us illustrate with the toy version of the dynamical law in which force is pretended to be constant and acceleration a scalar:

$$1 = kma.$$

k here is a constant, with numerical value that depends on the scale used for mass and acceleration. The key move is for the mixed absolutist to pick—arbitrarily!— one particular determinate mass property, m_0, and one particular determinate acceleration property, a_0, and use those to determine the value of k, given any chosen representation functions for mass and acceleration. The determination works as follows. Choose some initial numerical scales for mass and acceleration— say, ones in which m_0 and a_0 both have value 1. The constant k in the dynamical law has some particular numerical value v in this scale. Moreover, in any other scales, the numerical value of k is a function of the number v and the transformation constants relating the new scales for mass and acceleration to the old. This in turn means that there is a function of real numbers, $K_0(x, y)$, which yields the value of k in any scales for mass and acceleration as a function of the values of m_0 and a_0, respectively, in those scales (since the values for m_0 and a_0 in a scale determine the transformation constants between those scales and the initial scales). We can then write the dynamical law in terms of m_0, a_0, and this function K_0:

For any property mass- and acceleration-functions, M and A, with corresponding mass- and acceleration-functions m and a, and for any particle p,

$$1 = K_0(M(m_0), A(a_0)) \cdot m(p) \cdot a(p).$$

A similar account can be given of more complex laws. When applied to the (non-toy) laws of Newtonian gravitational theory, the result will be deterministic.

Of course, a related approach is available to the comparativist, in which the arbitrarily chosen items are concrete objects rather than properties. Laws based on arbitrarily chosen concrete objects (such as those proposed by Mach (1893) in response to Newton's rotating bucket) are widely rejected. Partly this is because it should be nomically contingent whether the chosen objects even exist; but surely it is also because of the intolerable arbitrariness of the selection of the chosen objects. But now, returning to the proposal in the case of the mixed absolutist, even though there is no corresponding concern about nomic contingency (since it does *not* seem nomically contingent that there are such properties as m_0 and

a_0), the arbitrariness in the selection of those properties remains problematic. Any two properties would do, so why those two?[49]

The arbitrariness could be made to go away: instead of a single law, stated in terms of m_0, a_0, and K_0, we could instead have, for each m and a, a law-relative-to m and a, with its own constant k. But this is a return to relativizing and nomic quotienting.

So far we have found no absolutist approach to the laws based on representation functions that secures determinism without arbitrariness or nomic quotienting. But what of Fieldian "intrinsic" laws?

Field's approach for constructing an intrinsic version of a numerical law (governing some scalar quantities) works in two steps. First, the numerical law must be replaced with a ratio version: an identity between ratios of the values of quantities for a pair of objects. For instance, in the case of the toy version of Newton's dynamical law:

$$1 = k\,m(x)\,a(x) \qquad\qquad \text{(for any } x)$$

the ratio version is:

$$\frac{m(x)}{m(y)} = \frac{a(y)}{a(x)} \qquad\qquad \text{(for any } x, y)$$

(note the disappearance of the constant k). Second, an intrinsic version of the latter statement may then be constructed, using techniques that Field describes.

Now, one option for the mixed absolutist would be to follow Field in step one—converting the original law to a ratio version—but then diverge from him in step two: the intrinsic version of the ratio version will now be a statement about the determinate properties and higher-order structuring relations, rather than a statement about the comparative relations over concrete objects.[50]

But this approach won't save indeterminism. For it retains the first of Field's steps, namely restating the usual law as a statement about ratios, but this makes the constants in the laws disappear.

In a little more detail, it would seem that the following is true:

---

[49] Martens (2017a, section 4.2.3) points out that one could cut down on the arbitrariness by choosing the mass property corresponding to the total mass in the universe.

[50] For example, $\frac{m(x)}{m(y)} = \frac{a(y)}{a(x)}$ can be given the following intrinsic statement, which is just like I-Newton from section 4.5 except that acceleration and mass sequences are now sequences of determinate properties, not concrete objects:

Absolutist I-Newton For any objects x and y, with determinate acceleration and mass properties a_x, a_y, m_x, m_y, there do *not* exist a mass sequence S_1 and an acceleration sequence S_2 such that (i) $m_x \in S_1$; (ii) $m_x \in S_2$; (iii) there are exactly as many members of S_1 that are $\leqslant_m m_x$ as there are members of S_2 that are $\leqslant_a m_y$; and (iv) there are fewer members of S_2 that are $\leqslant_a m_x$ than there are members of S_1 that are $\leqslant_m m_y$; and there do *not* exist an acceleration sequence S_1 and a mass sequence S_2 such that (i) $m_y \in S_1$; (ii) $m_y \in S_2$; (iii) there are exactly as many members of S_1 that are $\leqslant_a m_y$ as there are members of S_2 that are $\leqslant_m$ than m_x; and (iv) there are fewer members of S_2 that are $\leqslant_m m_y$ than there are members of S_1 that are $\leqslant_a m_x$.

Insensivity to mass-doubling For any mixed absolutist Fieldean version of
   Newton's laws, if a mixed-absolutist world obeys those laws then so does
   any other world that differs only by all masses being doubled.

For the first of Field's steps consists of replacing Newton's laws with ratio versions.
Since ratio versions will surely be preserved under mass doublings, any intrinsic
versions of them must likewise be preserved under mass doublings.

   That ratio versions are indeed preserved under mass doublings is clear in the
case of the toy version $1 = ma$ of Newton's dynamical law: if its ratio version
$\frac{m(x)}{m(y)} = \frac{a(y)}{a(x)}$ is true at some world, it will also be true when all objects' masses
are doubled (since $\frac{m(x)}{m(y)} = \frac{2m(x)}{2m(y)}$). To check this for non-toy laws of Newtonian
gravitation, we must look in detail at how Field's first step will go in that case. Field
treats mass density as a fundamental field, and replaces force with a fundamental
scalar gravitational potential field. His informal statements of the ratio forms of
the laws are these:

Law of gravitation

At any two points at which the mass density is not zero, the ratio of the Laplaceans of the
gravitational potential is equal to the ratio of the mass-densities. (1980, p. 79)

Dynamics

[T]he acceleration of a point-particle subject only to gravitational forces is at each point
on the particle's trajectory equal to the gradient of the gravitational potential at that point.
The invariant content of this law is exhausted by the claim that the gradient is proportional
to the acceleration . . .. (1980, p. 81)

Doubling the mass density field everywhere will preserve satisfaction of these two
laws. The dynamical law is clearly preserved since it doesn't mention mass-density
at all, and the law of gravitation is preserved since mass-density occurs only as a
ratio between mass-densities at a pair of points.

   I assume, then, that Insensitivity to mass-doubling is true. Now for the argument that intrinsic versions of Newton's laws of the sort under consideration—
mixed absolutist intrinsic versions of ratio-versions of the laws—will be indeterministic. Begin with the flat-footed absolutist worlds, A-Earth and A-Pandora,
considered in section 4.8 above. Next construct "mixed absolutist versions" of
these worlds:[51]

   MA-Earth: mixed absolutist version of A-Earth: nondoubled masses; projectile
      escapes; intrinsic mixed-absolutist Newtonian laws hold.
   MA-Pandora: mixed absolutist version of A-Pandora: doubled masses; projectile
      returns; intrinsic mixed-absolutist Newtonian laws hold.

Now consider a mixed-absolutist world, call it Doubled MA-Earth, that is just like
MA-Earth except that all masses are doubled. Since the intrinsic mixed-absolutist

[51] By assuming that such worlds can be constructed we are restricting our attention to conceptions
of the laws that obey a certain constraint, much as we did in our reformulation of Baker's argument in
section 4.8.

versions of Newton's laws hold in MA-Earth, by the principle of Insensitivity to mass-doubling they must also hold in this new world. Thus we have:

Doubled MA-Earth: exactly like MA-Earth but with doubled masses; projectile escapes; intrinsic mixed-absolutist Newtonian laws hold.

MA-Pandora and Doubled MA-Earth now show that the intrinsic mixed-absolutist Newtonian laws are indeterministic.

Thus Baker's problem of indeterminism arises even for absolutists who adopt intrinsic laws, if they follow Field's approach for constructing those intrinsic laws. For the first step of that approach obliterates the physical significance of the constants in the laws.

Verónica Gómez pointed out an absolutist strategy for restoring determinism: to an indeterministic law that secures only ratios of some quantity for pairs of objects, add a supplemental law which specifies the absolute value of that quantity in one particular case. This strategy could deliver deterministic intrinsic laws. For example, Field's strategy can be used (see note 50) to construct an intrinsic version of the ratio version $\frac{m(x)}{m(y)} = \frac{a(y)}{a(x)}$ of the toy Newtonian dynamics, and then an intrinsic supplemental law could be added specifying which absolute acceleration property would result from one arbitrarily chosen absolute mass property; and presumably such a strategy would work in non-toy cases. But the use of arbitrary entities (or avoiding this by relativization) would be as unappealing here as it was before.

4.11 Pessimistic conclusions

Our search in the previous section for an attractive conception of absolutist laws was as unsatisfactory as our earlier search for attractive comparativist laws. Whether we accept absolutism or comparativism our pick seems to be amongst poisons, the chief of which are: arbitrariness in the fundamental facts, nomic quotienting, and (Cthulhu forbid!) abandoning our overall approach to quantity based on the metaphysical tool of fundamentality.

Suppose we were willing to swallow a heavy dose of arbitrariness in the fundamental facts. We might then retrace our steps all the way back to flat-footed absolutism, according to which the fundamental quantitative concepts relate concrete objects to abstract entities such as real numbers. Each quantity would then have a metaphysically distinguished unit, which would be metaphysically arbitrary (to a degree that is hard to overstate). But we could have simple and attractive and deterministic (if this is empirically supported) laws. Or anyway we could have those if we also accepted what would seem to be further arbitrariness in our metaphysics of mathematics: sui generis real numbers and any other abstract entities involved in the fundamental quantitative laws.

Suppose instead that we could swallow nomic quotienting. We would thereby abandon the ideal of one set of laws to rule them all, one set of laws that would underly all formulations of the laws based on arbitrary choices. Then some options would open up. We could accept either absolutism or comparativism, depending on how poisonous we regarded the comparativist's commitment to the temporal nonlocality of the state of the world at a time. In either case the relativization approach could be taken: laws would be relativized to choices of representation functions, and a sort of determinism could be embraced (though the account of inter-world obedience would be kludgy in the comparativist case). Alternatively, the laws could taken to be intrinsic in Field's sense, with determinism secured by arbitrary choices. The concern was raised in section 4.7 that Field's approach is ultimately no less artificial than the representation-function alternative, since his laws quantify over extraneous entities and involve arbitrary and artificial choices. Setting aside whether these demerits are worse than the corresponding ones for representation-function laws, the concerns are answered by nomic quotienting. For the nomic quotienter has given up on the idea of explaining such artificialities in terms of the One True Laws. She is content to grant that what is nomically fixed about the world can be formulated in many different ways, each of which involves artificial or arbitrary choices: arbitrary choices of representation functions perhaps, artificial quantification over standard sequences, arbitrary choices of how to define a standard sequence or otherwise execute the Fieldian programme, and so forth.

Note that if the nomic quotienter does not want to swallow the first poison as well, she must accept a Humean or other reductionist metaphysics of laws. For if there were fundamental facts to the effect that the laws are such-and-such, then the artificial or arbitrary choices that the nomic quotienter embraces will have found their way into the fundamental facts themselves, not just into the laws.[52]

Finally, we might abandon the metametaphysical assumptions that have led us into this sorry situation. I have been assuming in this chapter that the right metaphysical tool for the metaphysics of quantity is fundamentality: that the way to articulate a metaphysics of quantity is to identify the fundamental quantitative concepts (or properties and relations), and that the best way to do so is to look for fundamental quantitative concepts that can enter into an attractive conception of the laws of nature. To my mind this metametaphysics had been the clear front-runner, coming into our discussion of quantity. (I think this in part because I accept much of the postmodalist critique of modal tools, in part because of how it has emerged in this book that ground and essence are inappropriate tools for articulating "low-level" metaphysical views, and in part because of my defence of the tools of fundamentality in Sider (2011).) But perhaps quantity is fundamentality's Waterloo.

Should we embrace immense arbitrariness in the most fundamental facts? (Surely not.) Should we reject the metametaphysics of fundamentality? (Some will say yes here, and though I don't say this is unreasonable I don't think it's our best alternative.) Or should we embrace nomic quotienting? (I don't see a

---

[52] See also note 44.

better alternative at present, though my real hope is for the future, for the menu to improve.) Regardless of our answers to these questions, we have yet another illustration of how the metaphysics of science is intertwined with the question of the conceptual tools with which the issues should be framed.

4.12 Appendix: ramsifying fundamental properties away

In section 4.4 I objected that dispensing with fundamental properties like mass precludes simple laws. But a variant strategy attempts to reinstate them. Instead of list-like laws specifying the allowable trajectories, one could instead "ramsify mass away", and have laws that claim that some property or other plays the mass-role in constraining trajectories. Simplify by pretending that 'mass' isn't quantitative, and is a simple monadic predicate.[53] Then replace $\mathscr{L}(\text{mass})$, the conjunction of the laws normally thought to govern 'mass', with $\exists p \mathscr{L}(p)$, which is neither infinitary nor list-like. And replace any statement $S(\text{mass})$ of a particular matter of fact about mass with $\exists p(\mathscr{L}(p) \wedge S(p))$.[54]

If the existentially quantified variable p in the ramsey sentence $\exists p \mathscr{L}(p)$ were restricted to "sparse" properties, then the only candidate value for that variable would be the property of mass itself, the very property we were trying to avoid, and no advance would have been made (unless we regressed back to the view that existentials need not be grounded by their instances). But the idea instead is to not restrict the variable in this way. The variable p is to range over properties in the abundant sense, which can be arbitrarily disjunctive, or even "nonqualitative", corresponding to sets with no defining condition. Such a law therefore says, in effect, that things can be divided into two groups, one of which behaves exactly as the class of massive objects is normally taken to behave. Contentiously put: those things behave *as if* they had mass—as if they shared some property in common that plays the mass role.

This approach has an inherent limitation. One might wish to eliminate more than one fundamental quantity. That would mean replacing the conjunction of the usual laws $\mathscr{L}(t_1,\ldots,t_n)$ governing the terms $t_1,\ldots,t_n$ for the quantities to be eliminated with the ramsey sentence $\exists p_1 \ldots \exists p_n \mathscr{L}(p_1,\ldots,p_n)$. But this ramsey sentence might be too weak a statement to be a law. It will certainly be too weak if $t_1,\ldots,t_n$ are *all* the nonlogical expressions in $\mathscr{L}(t_1,\ldots,t_n)$, for then the ramsey sentence will be guaranteed to be true provided $\mathscr{L}(t_1,\ldots,t_n)$ was consistent (and provided there exist sufficiently many objects). Thus ramsifiers do not try to eliminate absolutely all fundamental concepts from physics. In particular, they normally exempt spatiotemporal concepts.

---

[53] This pretence could be eliminated in various ways, depending on one's approach to quantities. The laws recognized by a comparativist, for instance, will take the form $\mathscr{L}(\succeq, C)$, and could be replaced with $\exists R^2 \exists R^3 \mathscr{L}(R^2, R^3)$.

[54] Compare Esfeld (2020); Hall (2015, section 5.2).

This approach is usually paired with the Humean view of laws. (Perhaps this is because many ramsifiers have modeled their approach on David Lewis's (1994) reduction of chance, as we'll see below.) But the approach isn't essentially tied to the Humean view; all that's needed is that ramsey sentences can be laws.

It would be productive for the ramsifiers to engage with Cian Dorr's critique:[55]

If we want to weaken a theory so as to eliminate its commitment to some sort of hidden structure, we can often do so by replacing the vocabulary which purports to characterize this structure with variables of an appropriate sort bound by initial existential quantifiers. Philosophers who are suspicious of particular putative bits of hidden structure keep on rediscovering this fact, and announcing that they have shown how to eliminate the structure in question. But once we have realized the complete generality of the trick, we should not be impressed by their achievements. Here are some more examples . . .

The examples Dorr goes on to list include eliminating fundamental kinds of particles, eliminating spatiotemporal structure above a certain level (say, topological) by quantification over coordinate systems, and (alas) my own four-dimensionalist account of the rotating homogeneous disk.

Another member can be added to Dorr's list: the recently popular proposal to understand the wave function in Bohmian mechanics as being determined by the entire history of particle positions. On this view, we replace the Bohmian laws $T(\Psi)$ about the wave function Ψ with $\exists\psi T(\psi)$, where ψ is a variable over functions from points of configuration space to complex numbers.[56]

According to Dorr, for certain members of the list—especially the first, eliminating fundamental kinds of particles—ramsification is clearly wrong-headed. Scientific realists should take it as a fixed point that ordinary empirical evidence can favour the conclusion that not all particles are intrinsically alike, and that not all spatiotemporally isomorphic objects are intrinsically alike. Thus we should reject the ramsification gambit in general. Ordinary empirical evidence favours laws that are stronger than the ramsey sentence; it favours laws containing non-logical constants for "hidden structure", such as 'charge', 'mass', and so on. The ramsey sentences are explanatorily worse than the unramsified sentences. Further, Dorr claims, we can point to a problematic feature, a distinctive theoretical vice, that accounts for the explanatory deficiency of the putative ramsified laws: the occurrence of initial existential quantifiers in them. (More on this later in this section.)

Dorr mentions a final target of his critique: ramsifying away all unobservable facts!

---

[55] Dorr (2010a, p. 160); see also Dorr (2007, section 3).

[56] Compare Bhogal and Perry (2017); Esfeld et al. (2014); Miller (2013). Yet another addition: in section 2.6 we saw that resemblance nominalism gives nomic essentialists the main thing they want: it blocks exchanging of nomic roles. But the resemblance nominalist can't state a law for charge, say, by name. The law must instead be that there exists some resemblance class that behaves in such-and-such a way.

We need only define up some notion of what it is for a model to 'accurately represent the observable facts': then our new theory can simply say that the old theory [which posited unobservable facts] is true in some model that accurately represents the observable facts. (p. 162)

If the ramsifier needed to admit this instance of her strategy, that would surely be fatal. However, our question here (unlike Dorr's) is that of which concepts are fundamental, and thus our ramsifier pursues her strategy only when the predicates that remain unramsified are fundamental. Since the concepts used to articulate "the observable facts" would need to be nonfundamental (by independent criteria— they wouldn't enable simple and strong laws, for instance), our ramsifier can give a principled reason for denying that her approach leads to eliminating all unobservables: 'the old theory is true in some model that accurately represents such-and-such observable facts' needn't be regarded as a law because it contains nonfundamental vocabulary.

Still, even though the death blow can be avoided, Dorr's objection to the ramsification strategy strikes me as serious. The ramsifiers owe him a reply.

The ramsifiers may be tempted to brazen it out, to claim that there is absolutely nothing wrong with any of the instances of the strategy that Dorr mentions (other than ramsifying away unobservable facts, which we have seen to be criticizable on independent grounds). I don't have a definitive objection, only a series of observations.

First, we shouldn't lose sight of how unintuitive ramsification is. 'This particle moved as it did because it's charged and all charged particles move that way' sounds like a great explanation; 'this particle moved as it did because it is one of a group of particles that all move that way' sounds like a horrible one.[57]

---

[57] Compare Dorr (2007, p. 39). It may be objected that even the first is a horrible explanation: the fact F_{all} that all charged particles move in way W cannot be part of the explanation of the fact F_a that a particular particle, a, moved in way W, since a universal generalization F_{all} is partly explained by its "instances" such as F_a. This is like a familiar objection made by antiHumeans that Humean laws cannot explain. Loewer (2007, p. 321) has defended Humean laws from the objection, and we can follow him in replying to the present objection: although F_a *metaphysically* explains F_{all}, F_{all} *scientifically* explains F_a. I am a fan of this Loewerian response; the argument in the text is meant to appeal to other fans (although an antiHumean could make a related argument). Talk of explanation in the text is intended in the scientific sense; the argument thus consists of these two claims:

(1) All charged particles moving in way W, plus particle a being charged, *does* scientifically explain particle a moving in way W.

(2) There being some particles that include a and all move in way W does *not* scientifically explain particle a moving in way W.

A ramsifier might reply that statements about explanation must themselves be submitted to ramsification. (Similarly, a statement of determinism would need to be submitted to ramsification.) Thus even conceding (2), the ramsifier might claim that:

(3) There are some particles such that (i) each of those particles moves in way W, and (ii) a being one of those particles scientifically explains its moving in way W.

I myself find even (3) objectionable, but perhaps the ramsifier will claim to be in an improved position.

Second, the analogy between existential quantification and disjunction might give us pause, insofar as we regard disjunction as anethema to lawhood. The ramsifier's law $\exists p \mathscr{L}(p)$ is akin to $\mathscr{L}(P_1) \lor \mathscr{L}(P_2) \lor \ldots$.

Third, the fact that the ramsification strategy can't be pursued "all the way" might be regarded as having various kinds of significance. To avoid trivialization, some "constants" must be retained, to populate the ramsey sentence. But then the concerns with the old theoretical terms that led to ramsifying them away might apply to the retained constants. (This wouldn't always be the case, not if the concerns are simply those of ideological parsimony, or if they were specific to the ramsified-out terms, as in the case of the wave function.) Also, a distinction between first- and second-class laws would result, the second-class ones involving ramsification and the first-class ones involving only the constant terms but no ramsification, as well as a correlative distinction between first- and second-class explanation.

Fourth, trivial truth isn't the only potential danger of ramsifying too much. In classical physics, suppose one claimed that the only fundamental structure was the following: there exist points of time and points of space and particles; there is a fundamental relation of occupation that particles bear to pairs of points of space and time (so that particle trajectories are fundamental); space and time have only topological fundamental structure. Thus there is no fundamental metric, no fundamental electromagnetic field, and no fundamental charges or masses. The laws are existential in form: particle trajectories behave, by law, as if there exists a metric on space and on time, as if there is an electromagnetic field on space over time, and as if there are particle charges and masses, which all constrain the trajectories as classical physics says. Even if these existential sentences aren't so weak as to be trivially true (provided the domain is large enough), they may be weak to a lesser extent that creates other problems. If, for instance, they don't constrain trajectories any more than, say, to continuous trajectories, then they may not be the laws (or anyway, we may not be justified in thinking they're the laws) because the more simple intrinsic statement 'objects move continuously' would be a better candidate for lawhood. Another possibility: the weakening of the law might make it indeterministic.[58]

Fifth, combinations of acceptable instances of the strategy can be, in aggregate, unacceptable. It can happen that ramsifying out either of two aspects while leaving the other unramsified keeps the laws reasonably strong, whereas ramsifying both would not. One worries that this is a sign that the ramsification strategy in general is intrinsically problematic.

Sixth, a concern that is specific to ramsification of the wave function: since configuration space is not taken to itself be physically fundamental by the ramsification approach, this introduces another degree of freedom that moves the

---

[58] It might be objected that the entire statement of determinism ought to be ramsified; compare the end of note 57.

ramsified laws one step closer to triviality. Even if one is happy with the ramsification approach in the case of, say, eliminating particle kinds—happy, that is,
with laws saying in effect that particles behave as if they come in a small number
of different kinds which have distinctive sorts of trajectories—one may not be
happy with a second layer of "as if", with laws saying that particles behave as if
there is a fundamental space with the geometry of configuration space and as
if there is a field on that space, which field constrains trajectories in ordinary
three-dimensional space. However, as Eddy Chen pointed out to me, this concern
could be avoided by ramsifying out a "multi-field" (Forrest, 1988; Belot, 2012;
Chen, 2017) on physical space rather than a complex field on configuration space.

There are two further cases in which Dorr's critique might seem to apply;
but in the end I think that it does not.[59] The first is what David Albert (2000)
calls the 'past hypothesis', a law saying that *at some* point in time, the universe was
in a very low-entropy state. Dorr's critique seems at first to apply, since the past
hypothesis is existential in form.[60] However, the intuitive explanatory badness that
Dorr abhors does not seem present here. Ramsifying away mass and charge in
classical physics, for example, seems bad because it is an artificial weakening of an
explanatorily superior theory—one is in effect saying that objects' motions are as if
they have masses and charges. But the past hypothesis doesn't seem like an artificial
weakening of any stronger claim that presupposes more structure. It is offered
as a solution to the problem of accounting for temporally asymmetric macro-
phenomena in a world in which the fundamental dynamical laws are insensitive
to temporal direction; and there would not seem to be any stronger version of
the past hypothesis that solves the problem but embraces more structure. (If the

---

[59] Yet another is relativization in the sense of section 4.9. Why might one think that Dorr's critique
applies? After all, the relativizer's laws involve no existential quantification. (The ramsifier about mass,
for instance, can pick out the class of objects with a certain mass only by the role that those objects
play in the laws of motion; that is why existential quantification is needed to formulate those laws of
motion. The relativizer, on the other hand, can pick out the class of mass-functions without appealing
to the laws of motion—by using the fundamental comparative predicates—and thus can formulate the
laws of motion without existential quantification.) The argument would need to be that relativized
laws are an artificial weakening of absolutist laws, and as a result are explanatorily inferior to them.
Now, maybe they are explanatorily inferior, but if they are, it is for reasons quite different from why
paradigmatic ramsificationist explanations, such as 'it moved thus because it is one of a class of things
that all move thus', are inferior. The reasons would rather be those discussed above: perhaps the
nonlocality of relativization, or its difficulty in accounting for inter-world obedience, or—to my mind
the most serious—their basis in arbitrary decisions. Also, as argued in section 4.10, it is unclear whether
there are any better absolutist alternatives.

[60] Objection: its form should instead be: 'the past temporal boundary of the universe has low
entropy'. (This would require antireductionism about time's arrow. Although most advocates of the
past hypothesis are reductionists about time's arrow, an antireductionist could advocate it as well.)
Reply 1: to pick a nit, this is still existential given Russell's (1905) approach to the first word of the
proposed law. Reply 2: more importantly, one might wish the past hypothesis to not require that the
initial low-entropy state be at the very beginning of time, but merely at the beginning of the portion
to which we have epistemic access (Albert, 2000, p. 85, note 13), in which case it should say merely
that there is some very low-entropy state.

fundamental dynamical laws are time-reversal symmetric, positing a fundamental direction of time doesn't on its own solve the problem.)

If this is correct, does it call into question Dorr's diagnosis of the problem, his claim that the distinctive source of explanatory badness in the paradigm cases of ramsification is initial existential quantifiers? Perhaps not, for his claim is actually more subtle than that (I have been oversimplifying). Dorr's claim is not a blanket prohibition of initial existential quantification; rather, the explanatory badness associated with an initial existential quantifier comes in degrees, and increases with, roughly, the amount of the theory that needs to be governed by that quantifier:

Consider for example a theory that says that there is a point of space towards which all bodies accelerate (in certain specified ways). It seems to me that if we found out that there was such a point, we would have reason to think that it was intrinsically special, *or at least that it could be distinguished by some structural role simpler than that of being a point towards which bodies accelerate in the specified ways.*

(Dorr, 2010a, p. 163, my italics)

... one can improve a theory by replacing a long existential quantification '$\exists x\,\phi(x)$' with a conjunction of the form '$\exists x\,\psi(x) \wedge \forall x(\psi(x) \rightarrow \phi(x))$', where ψ is considerably shorter than ϕ.

(Dorr, 2010a, p. 166)

This way of understanding his diagnosis perhaps exonerates the past hypothesis. An existential quantifier need only govern a single facet of the laws, the past hypothesis; it needn't govern any of the fundamental dynamical laws or any laws about nondynamical chances. In this way it is unlike the existential quantification in ramsified laws about charge and mass, which must govern all reference to charge and mass in the laws (and in particular matters of fact as well). Perhaps, then, the existential quantification in the past hypothesis is a teensy bit bad, but the badness is outweighed by other explanatory virtues; and there is no competing hypothesis that is explanatorily superior.

The second further case in which Dorr's critique might seem to apply is Lewis's theory of chance itself. Now, this suggestion might seem odd. After all, Lewis does not present his reduction of chances as an instance of ramsification. His approach is instead the following. He begins with an undefined predicate $Ch(P, t, x)$ of propositions P, times t, and real numbers x. He then considers various partially interpreted theories employing 'Ch' as well as predicates for natural properties. Of these partially interpreted theories, setting aside the ones that imply false non-'Ch'-involving statements, the rest are put into a competition in which the goal is to optimally (under some measure) combine three virtues: syntactic simplicity (of a certain sort), the amount of truth implied about non-'Ch'-involving matters (under a suitable measure), and "fit", a measure of how well the theory's "chances"—that is, the numbers that, according to the theory, are associated by Ch with propositions and times—match the frequencies. The theory that wins this competition—the "best system"—is said to give the truth about both chance and the laws. In this presentation there is no overt ramsification.

But the presentation is unspecific in a crucial way. No account is given of what the laws of chance $\mathscr{L}(\mathrm{Ch})$ in the winning theory *say*, or of what statements $A(\mathrm{Ch})$ of particular matters of fact about chance say. (The account, for instance, does not fix the modal profiles of any of these sentences.) All that is given is an account of the *extension* of Ch, and thus of the *truth values* of statements of chance. And the obvious way to fill this gap is ramsification. What the laws of chance say is that $\exists c \mathscr{L}(c)$; what statements of particulars of matters of fact about chance say is $\exists c(\mathscr{L}(c) \wedge A(c))$.

The only other way to fill the gap would seem to be purely "extensional": one could regard Ch as naming the relation, C, between propositions, times, and real numbers such that $\mathscr{L}(C)$ in fact best combines simplicity, strength, and fit.[61] But this would give the laws the wrong modal profile, and would make them extraordinarily complex (since there is no guarantee that C will be definable using predicates for fundamental properties and relations—in general it will be a mathematical function "given in extension"). More importantly, interpreting predicates for hidden structure in this extensional way is available to all of Dorr's targets. If his critique is on track, it must apply to the extensional variants as well as ramsification.[62]

When rightly understood, then, Lewis's account of chance involves ramsification. Thus one might regard Lewis as ramsifying away hidden structure that ought to be retained: a fundamental quantity of chance.[63] And yet, as with the past hypothesis, I am not inclined to follow the argument where it leads here, since Lewis's account of chance does not seem like an artificial weakening of some

---

[61] It may be argued that there is a third way to fill the gap: say that the constraints put on the extension of Ch secure it a nonextensional and nonramsifying interpretation, even though no such interpretation has been specified. This might be coupled with heavy reliance on grounding, by which I mean an appeal to a realm of nonfundamental facts that resist definition (in any sense) in terms of fundamental facts, but rather are asserted to be connected to the fundamental facts by means of an undefined grounding relation. (Does p. 78 of Bhogal and Perry (2017) suggest such an idea, in the case of the wave function?) In my view, this is just a metaphysically unspecific claim, with the first two ways of filling the gap the only available ways of making it specific. This criticism is in the spirit of section 2.3.

[62] As mentioned in note 56, those who reduce the Bohmian wave function to particle positions do not explicitly adopt ramsification. That is because they model their reductions on Lewis's reduction of chance. But their accounts inherit a corresponding unspecificity: they do not specify what statements about the wave function say. As with Lewis, the best way to fill that gap is ramsification. (The mitigating considerations in the case of Lewis that I am about to discuss do not let them off the hook.)

[63] Lewis objected that primitivism would make a mystery of the connection between chance and credence (1994, pp. 484–5). But a primitivist could adopt Lewis's own explanation of this connection, by appealing to the fundamental quantity's lawful connection to frequencies; that quantity would neither constrain credence nor deserve the name 'chance' in worlds in which this connection is absent. This approach admits certain possibilities for global scientific error that Lewis's approach does not, in worlds in which the primitive quantity doesn't match what Lewis would call the chances. But anyone who posits fundamental quantities must admit such possibilities of error. If charge and mass are fundamental quantities posited to explain the motions of particles, there are possible worlds in which particles move as if they have those quantities whereas in fact they do not.

stronger and explanatorily superior theory. The putatively stronger and superior theory would be one based on primitive chances; but that theory is not explanatorily superior, I think, because the primitive chances would constitute explanatorily idle structure. With Lewisian chances, stochastic laws look in essence like this: 'the frequencies of F_1s that are G_1s is approximately and pervasively the number x_1, the frequencies of F_2s that are G_2s is approximately and pervasively the number x_2, …'. With primitive chances one is adding that there is a fundamental quantity C that mediates all of these statistical correlations. The laws thus bifurcate; instead of proceeding directly to frequencies, they say, first, that there is some fact involving C, F_1, G_1, and the number x_1, and another fact involving C, F_2, G_2, and the number x_2, etc.; and then they say, second, that given the first fact, the frequency of F_1s that are G_1s is, pervasively and approximately, x_1, and that given the second fact, the frequency of F_2s that are G_2s is, pervasively and approximately, x_2, and so on. The laws of the second approach add nothing to explanations but a wheel that idly turns.

It might be objected that charge and mass are also wheels that idly turn. Michael Esfeld (2020, section 3) says exactly this:

The argument for this claim is the one illustrated in Molière's piece *Le malade imaginaire*: one does not explain why people fall asleep after the consumption of opium by subscribing to an ontological commitment to a dormitive virtue of opium, because that dormitive virtue is *defined* in terms of its functional role to make people fall asleep after the consumption of opium. By the same token, one does not obtain a gain in explaining attractive particle motion by subscribing to an ontological commitment to gravitational mass as a property of the particles, because mass is *defined* in terms of its functional role of making objects attract one another as described by the law of gravitation.

But that would be to treat 'wheels that idly turn' as far too indiscriminate a critique. It surely *is* explanatory to say that there are intrinsic differences between subatomic particles that explain why they move differently. (As an antireductionist about geometric concepts, even Esfeld must admit that it is explanatory to posit fundamental geometric concepts to explain what we perceive.) The complaint about primitive chances is more specific: that once we have posited the necessary hidden structure ($F_1, G_1, F_2, G_2, \ldots$ in the previous paragraph) to characterize the stable approximate frequencies, positing C as *further* interposed structure does not improve the explanation. This complaint is exactly like the complaint that primitive lawhood adds nothing to explanations: 'B because: (i) A and (ii) it's a primitive law that if A then B' is no better than 'B because A'. To be sure, some (such as Armstrong (1983, chapter 4)) will say that primitive lawhood *does* add to the explanation. But even they will agree that adding a further turning wheel to the mix, a notion of primitive meta-lawhood, which mediates the connection between primitive lawhood and regularities, would not improve the explanation. There is a substantive and difficult question here of when posits improve explanations. Though I do not have a general answer to that question, it is a sensible view, I say, that positing fundamental charge and mass does improve the explanation, whereas positing fundamental lawhood and chance does not.

If positing charge and mass is indeed explanatory, then how is it unlike positing a dormitive virtue? What is wrong with Esfeld's argument? First a preliminary point: mass is not *defined* by its functional role (nor is charge). Mass is rather *posited* as a fundamental (and thus incapable of definition) property which in fact plays that role, a role in the fundamental laws of motion. The functional role may fix the reference of 'mass', but it does not supply a synonym (or a modally equivalent specification). Of course, one could say the same thing about the dormitive virtue. But Molière's doctors were presumably not doing this—not positing a fundamental property and a fundamental law involving it governing sleep. *That* would be like a property dualist positing fundamental mental properties and fundamental laws governing them. Although I and other physicalists believe that doing so would be misguided (since we think that physics can ultimately explain mentality), and even more misguided in the case of sleep, it is not absurd. Rather, the doctors were simply giving a name, 'dormitive virtue', to whatever causes sleep—where 'whatever causes sleep' does not signal a new posited fundamental property or law, but rather accepts the status quo of physiological structure, and picks out whatever it is in that structure that is responsible for sleep. That *is* absurd. It adds absolutely nothing to what was already known: something about opium causes sleep.

5

Equivalence

Equivalent theories (or statements, or models, or representations) represent the very same state of the world; any differences are merely conventional or notational.[1] Our question is the metaphysics of equivalence, what it is to be equivalent. This question is intertwined with the previous chapters, as we will see.

I will focus on two opposing approaches. They are in a sense the extremes, and some may prefer some intermediate. But if intermediate approaches are unsatisfying, as they often are, we will have a stark choice before us.

5.1 Symmetry, translation, meaning, modality, grain

A symmetry (of the laws of nature) is a one-one mapping from the set of possible histories onto itself that never maps a solution of (or history compatible with) the laws to a nonsolution, or a nonsolution to a solution. The velocity boosts discussed in section 3.5, for instance, are symmetries of the laws of Newtonian gravitational theory. Symmetries are often invoked in discussions of equivalence by philosophers of physics. Jenann Ismael and Bas van Fraassen (2003), for example, say that if some symmetry maps one history to another, and if the histories are also perceptually indistinguishable (in a certain sense), then we have reason to think that the histories are equivalent. Others have discussed other claims of the same form, the form being: 'we have reason to think that (descriptions of) histories are equivalent if they are (i) related by a symmetry, and (ii) X'.

Claims of this form address the *epistemology* of equivalence, which appears to be the main issue addressed in the literature. And symmetries surely do play some sort of central role here. They are "beacons of redundancy", as Ismael and van Fraassen (2003, p. 391) say.

But our topic is what equivalence *is*, and I doubt very much that the answer to this question will take the form 'symmetry plus X'. A symmetry might map a history to an utterly dissimilar history, as long as it never associates solutions with nonsolutions.[2] The history that actually occurs, for example, might get mapped to a history in which there is no sentient life at all, or a history in which only a

---

[1] When I say 'the *very same* state of the world', I mean it. Agreeing on just one aspect—for example, matters that are observationally accessible to us—isn't sufficient if there are differences regarding other genuine aspects of reality.

[2] This is a commonly made point; see Ismael and van Fraassen (2003); Belot (2013).

single particle exists. This is why an X always gets added to symmetry/equivalence principles; but no such X, it seems to me, could turn symmetry into an account of what equivalence is. Since histories related by a symmetry can be as dissimilar as you like, the concept of a symmetry doesn't even *approach* the concept of equivalence—it isn't as if symmetry is "almost" equivalence, or a "part" of equivalence. And given this, it's hard to see how adding some further condition X would get us closer to the concept of equivalence, unless X amounts to equivalence on its own.

Also, some added Xs prejudge certain issues. For example, requiring the symmetry to be continuous with respect to the topology of the space of histories just bakes in that topological structure isn't conventional.[3] This is a plausible claim, but it shouldn't be settled by the very definition of equivalence. (For that matter, *any* 'symmetry plus X' definition bakes in the objectivity of the laws.) To be sure, such objections are delicate, since any contentful account of equivalence will bake in something. The objection rests on the assumption that these particular claims about equivalence shouldn't be prejudged.

Symmetry is presumably a necessary condition for equivalence[4] (assuming that the laws are indeed objective), since the laws are presumably sensitive only to genuine features of reality, not conventional artefacts of representations of reality. Moreover, other things being equal we ought to avoid recognizing genuine features of fundamental reality that physics doesn't care about. These observations speak in favour of symmetry as a guide, albeit a defeasible one, to equivalence. But if our interest is in what equivalence is, we must set symmetry aside.

Another approach that is popular in the philosophy of physics understands equivalence in terms of relations of translation, understood in a purely formal or syntactic way. Thomas Barrett and Hans Halvorson have recently distinguished various proposals in this vicinity, one due to W. V. O. Quine, one due to Clark Glymour, and a third due to Barrett and Halvorson themselves.[5] In Quine's version of the idea, the relation of translation holds when the predicates of either theory can be "reconstrued" in terms of the other theory; in Glymour's it holds when the theories share a "definitional extension"; and according to the third, it holds when the theories are "Morita equivalent", which is like sharing a definitional extension except that the extension can involve the introduction of new sorts in a many-sorted language, and new quantifiers, variables, and nonlogical symbols of the new sorts, in addition to new predicates, names, and function symbols of the old sorts.

---

[3] Via the way in which the topology of the space of histories is defined in terms of the topology of physical space-time (or some other physical manifold).

[4] This is consistent with there being no X; equivalence needn't "factorize". Compare Williamson (2000) on knowledge.

[5] Barrett and Halvorson (2016a; b; 2017); Glymour (1970; 1977); Quine (1975). Similar remarks apply to approaches based on category theory (Barrett and Halvorson, 2016b; Weatherall, 2016).

Whether two theories stand in any of these relations depends only on syntactic and logical features of the theories; what the nonlogical symbols in the theories actually mean plays no role whatsoever. Thus theories whose sentences have the same logical forms are guaranteed to stand in the relations. For instance, a theory whose sole axiom is 'Some molecule is part of some electron' bears all three of the relations to a theory whose sole axiom is 'Some electron is part of some molecule'. (In the case of Quine, this is because 'electron' can be reconstrued as 'molecule' and 'molecule' as 'electron'.) But the statement that some molecule is part of some electron is obviously not equivalent to the statement that some electron is part of some molecule—the statements differ in truth value, after all. What is going on?

It is clear what was going on for Quine. Quine was not considering the question of equivalence in full generality. That is, he was not considering the question of what it is for an arbitrarily chosen theory—articulated in some interpreted language, in some possible circumstance—to be equivalent to another such arbitrarily chosen theory—perhaps in a different language, perhaps in other circumstances. First, Quine was only considering theories that were put forward as the total theory of some person. Second, he was only considering pairs of theories that were put forward in the very same circumstances: in the same possible world, time, and place. And third, he was assuming that the people putting forward the theories meant the same things by their observational vocabulary.[6] Only for such theories did Quine say that reconstrual amounts to equivalence. So although Quine would concede that 'Some molecule is part of some electron' and 'Some electron is part of some molecule' are not equivalent, given how we actually use those sentences, he would say that if a member of some other linguistic community in the same actual circumstances as we are in were to put forward, as her total theory, a theory just like our total theory except with 'electron' and 'molecule' swapping their places throughout the theory, then her total theory would be equivalent to ours. For she would mean by 'electron' what we mean by 'molecule', and would mean by 'molecule' what we mean by 'electron' (p. 319). Quine is surely right about this.

It is important to appreciate how Quine's thesis mixes together metasemantics—that is, radical interpretation, the factors that determine meaning—with equivalence. It would be better to separate these aspects—to have a theory solely about equivalence.

The restrictions on the pairs of theories that fall under the scope of Quine's account are drastic. His account says nothing about equivalence between theories or statements that are proper parts of total theories, or are merely considered

---

[6] Quine doesn't put it this way, since he is taking pains to avoid talking about meaning. Instead he restricts his attention to theories in our own language, and considers reconstruals that are restricted to the theoretical vocabulary, the idea being that the interpretation of the observational vocabulary will thereby be the same without needing to speak of sameness of meaning. But it is then confusing why he considers the theories in which 'electron' and 'molecule' are permuted to be equivalent. He seems to shift from considering the permuted theory as we actually understand it to considering it as it would be understood if we had put it forward.

rather than put forward, or are located in different circumstances. An account solely about equivalence, which aspires to tell us what equivalence is, ought to cover all such cases. It should tell us, for example, that our sentences 'Some molecule is part of some electron' and 'Some electron is part of some molecule' are not equivalent.

Stepping back: the problem with defining equivalence as translatability in a purely formal sense is, intuitively, that nothing is said about the contents of the "translated" sentences.[7] The translated sentences ought additionally to be required to have the same contents, one wants to say; but specifying the relevant sort of sameness of content is difficult, and it's exactly what is at issue when we ask what equivalence is. Quine could bypass such difficult questions of sameness and difference in content, and make do with a purely formal conception of translatability, only because of the restricted scope of his account.[8] One we broaden the scope of the account, the difficult questions must be confronted, for without the restrictions the purely formal approach is a nonstarter.

Moreover, the metasemantics on which Quine relies is contentious. Let T_1 and T_2 be theories in the scope of Quine's approach that differ by a reconstrual, R. Each sentence S of T_1 is associated by a "translation" $R(S)$ in T_2, the translation function $R()$ being induced by R's reconstruals of T_1's atomic predicates; and the appeal of the claim that T_1 is equivalent to T_2 is dependent on the sentences in T_1 meaning the same, in some good sense, as their translations in T_2. Quine is effectively assuming a metasemantics that guarantees this.

But consider David Lewis's (1984) "reference magnetism", a metasemantics developed in response to the model-theoretic argument against realism. According to reference magnetism, the assignments to atomic predicates by a *correct*, or *intended* interpretation of a language must be properties that are as fundamental (natural, in Lewis's terms) as possible, in addition to satisfying any other constraints on interpretation (such as making an appropriate part of a speaker's theory come out true). Now, if a sentence S of T_1 and its translation $R(S)$ in T_2 have different truth values under their intended interpretations, then they surely do not mean the same in any good sense. But given reference magnetism, they might well differ in truth value. A Quinean reconstrual of an atomic predicate of T_1 need not be an atomic predicate of T_2; it can in general be any formula in T_2 with one free variable. If R reconstrues the atomic predicates of T_1 as sufficiently complex formulas of T_2, an interpretation might need to assign quite nonfundamental properties to the atomic predicates of T_2 in order to make each $R(S)$ have the same truth value as S, and thus might not count as an intended interpretation given reference magnetism.

---

[7] See also Sklar (1982, pp. 92–3).

[8] This is not a criticism of Quine (1975), whose goal was not to give an account of equivalence in general, but rather to show that cases of the underdetermination of theory by data are difficult to come by. All he needs to achieve this goal is a claim about equivalence with the restricted scope.

Thus Quine's account of equivalence is undermined by at least one approach to metasemantics—and a plausible one at that, in my view (Sider, 2011, section 3.2). Moreover, reference magnetism brings out a more general problem, which is that Quine is effectively relying on the assumption that the correct metasemantics is insensitive to differences between theoretical terms and their Quinean reconstruals. Reference magnetism is just one model on which this assumption is incorrect. There are other such models, for instance that causal relations to the environment—even for theoretical terms—are metasemantically relevant.

For these reasons, I don't think that translatability, understood in a purely formal way, can be the whole story about equivalence, and I will not consider it further. Nevertheless, the recent work on translatability has a lot to teach us. For one thing, translatability, in various senses, is an important part of the epistemology of equivalence. Formal parallelisms between theories, particularly of the kinds that physicists actually investigate, are powerfully suggestive of equivalence, and invite further inquiry into whether the parallelism is due to the theories actually getting at the same portion of reality. Moreover, if translatability is understood in a suitably general way,[9] it may well be a necessary condition on equivalence. Finally, I suspect that something like the formal accounts of translatability should be incorporated into any fully fleshed-out metaphysics of equivalence. For instance, my own proposed account will appeal to notions like ground, which in my view ought ultimately to be replaced with something more like translatability in some rigorously defined sense.

Purely formal accounts fail because they entirely neglect meaning. In light of this it is natural to wonder whether equivalence can be understood solely in terms of meaning. Why not say simply that theories are equivalent when they mean the same thing?

The notion of meaning the same is not especially clear, but that isn't my objection. (I suspect that any reasonable account of equivalence will have some such unclarity.) My main objection is rather that the ordinary notion of meaning is too fine-grained to be appropriate in the present context. Consider, for example, the claim that the Hamiltonian formulation of classical mechanics is equivalent to the traditional formulation. This may or may not be true, but it is certainly a claim of equivalence that has often been made in the philosophy of physics. But these formulations don't *mean* the same, not in any ordinary sense of meaning anyway. The target notion of equivalence is not a matter of ordinary meaning, since statements can be equivalent in the target sense, get at the same portion of reality, even if ordinary language knows nothing of this fact, so to speak.[10] Relatedly, on some views about meaning, structurally different sentences thereby have different meanings. The self-conjunction $A \wedge A$, for instance, has a different

---

[9] See Barrett and Halvorson (2017) for a nice example of the need for this.
[10] See also McSweeney (2017, pp. 270–1) and Sklar (1982, p. 94).

meaning from A.[11] But whether such views are correct is surely irrelevant to theoretical equivalence.

According to some, the meaning of a sentence is the set of possible worlds in which it is true (Stalnaker, 1984). This view denies that structurally different sentences express different propositions, and indeed might be thought to identify the propositions expressed by the Hamiltonian and standard formulations of classical mechanics. This might suggest a modal account of equivalence, on which theories are equivalent when they are true in exactly the same possible worlds. (It does not *demand* the modal account—we cannot simply assume that sameness of proposition, in the sense that is relevant to the philosophy of language, is relevant to theoretical equivalence.) But a mathematical theory is not equivalent to a theory consisting purely of logical tautologies, even if each theory is true in all worlds. Nor is a theory of the mixed absolutist's structuring relations over the determinate quantities (section 4.7.3) equivalent to the tautological theory, even if it too holds in all worlds. And even when the modal account of equivalence delivers the correct results, this may be for deeper postmodal reasons. We have reprised themes from Chapter 1: modality is too crude and superficial a tool for our purposes.

Throughout this book I have been writing as if the postmodal revolution is the very latest development in metaphysics. But actually there is a newer kid on the block, "higher order metaphysics", the exploration of metaphysical issues statable in languages with irreducibly higher-order quantification of various sorts.[12] A great deal of exciting work in this area is happening, and some of it might be thought to bear on the question of equivalence. For instance, in an appropriate higher-order language one can define a binary sentence operator $\equiv$ which stands, intuitively, for propositional identity (Goodman, 2017):

$$A \equiv B =_{df} \forall X (XA \leftrightarrow XB)$$

(X is a variable occupying the syntactic position of a monadic sentence operator.) One could then use this connective in an account of equivalence (in the case of individual sentences anyway): sentences A and B are equivalent if and only if $\ulcorner A \equiv B \urcorner$ is true. This raises many issues, about which I will say only the following (I hope to say more in the future). Suppose for the sake of argument that the higher-order languages are in good standing (a nontrivial assumption). Statements in those languages are truth-apt, objective, determinate, substantive, and so on, let us assume. Thus reality has a distinguished sort of "grain"—a distinguished notion of propositional content, so to speak. Even granting all this, it in no way follows that this distinguished grain has anything to do with theoretical equivalence. Judging from the literature, the considerations thought to be relevant to "grain science" seem quite distant from considerations in the philosophy of science or metaphysics thought relevant to theoretical equivalence. We would need an

---

[11] See King (2019) for an overview.
[12] See for instance Dorr (2016); Williamson (2013).

argument, anyway, that the grain-scientist's grain is of the right sort to be relevant to theoretical equivalence.[13]

Let us leave symmetry, translation, meaning, modality, and grain behind, turn to postmodal metaphysics, and directly investigate the concept of 'saying the same thing about the world'.

5.2 Examples: quantities, metric, ontology

We all can agree, surely, that theories differing only over units of measure are equivalent. What is it that we are all agreeing on?

Well, imagine someone who denies the claim. Imagine a person, "Kilo", who thinks that there is a "distinguished" unit of mass—kilograms, say. Just as some people think there is a distinguished direction of time, Kilo thinks that the kilogram unit is physically distinguished.

Kilo has perfectly ordinary views about what masses things have in any given scale. Kilo understands that statements about mass in grams are *true*, and takes them to be correlated with statements about mass in other scales in the same way that we all do. But a description in terms of grams isn't equivalent, Kilo says, to a description in terms of kilograms, because there are some *extra mass facts* that are left out by the grams description. If you say only that something is 1 kg, and that it is 1,000 g, you've left out that it's *really 1* in mass; it's 1 in the *distinguished* unit. But how should we think about this alleged "extra fact"? What would it mean for a unit to be "distinguished"?

Before trying to answer, consider two other cases in which parallel issues arise. First consider someone like Reichenbach (1958, chapter 1), who thought that certain theories making apparently incompatible claims about the curvature of space are in fact not incompatible, and can even be equivalent, if the predicates for talking about distance in the two theories are governed by different stipulated "coordinative definitions", which definitionally connect those predicates to other theoretical predicates (like 'force') and to procedures of measurement. The opposing "realist" view about distance says that spatial predicates need no coordinative definitions, but rather refer to the intrinsic geometry of space, which includes "distinguished" distance structure. Realism about the metric is the analogue of Kilo's view about units of mass, but unlike Kilo's view it isn't strange at all; indeed, it appears to be the majority view nowadays. But again: what does this talk of being "distinguished" amount to?

---

[13] For instance, I argue in the next section that various claims of equivalence turn on questions of fundamentality. Will the answers to those questions be reflected in facts about grain? Nothing in the grain science literature hints that this will be so. Indeed, one gets the impression that the higher-orderists think of their project as a rival to (what I have been calling) postmodal metaphysics. (Fritz (2021) argues powerfully that standard assumptions about ground clash with the higher-order framework, though I don't know of any clash with fundamentality per se.) There is much to investigate here.

Second, consider Eli Hirsch's (2011) views on (meta)ontology. There is a dispute amongst metaphysicians over whether there exist composite objects, objects with smaller parts. David Lewis (1986a, pp. 212–13) said yes: there really do exist objects with parts, such as molecules, chairs, planets, and so on. (Indeed, according to Lewis, there exist arbitrarily scattered objects, since for any objects whatsoever, there exists a mereological sum of those objects.) Peter van Inwagen (1990) said no, molecules, chairs, and planets do not exist. All that really exist are their subatomic parts.[14] Then along came Hirsch, who said: this is a nonissue! According to him (and really, lots of people, especially nonmetaphysicians), the following two claims are equivalent:

van Inwagen: there exist subatomic particles in a certain "chairlike" arrangement C; there does not exist any further object (a chair) containing those particles as parts.

Lewis: there exist subatomic particles in arrangement C, and there also exists a further object (namely, the chair) containing those subatomic particles as parts.

Hirsch's reason was, in essence, that van Inwagen and Lewis can each use his concept of existence (his existential quantifier) to define up a concept that fits the other's theory, and that there is no "distinguished" concept of existence. For example, van Inwagen can define up a new meaning of 'there are'—call it 'existence$_{Lewis}$'—under which he can agree that Lewis's claims come out true. He would simply need to define 'there are$_{Lewis}$ chairs' as meaning what he (van Inwagen) ordinarily would mean by 'there are subatomic particles in arrangement C'.

What would it take to oppose Hirsch and regard the ontological debate as being genuine after all? A distinguished concept of existence. Then even though everyone can agree on what "exists$_{Lewis}$" and what "exists$_{van\ Inwagen}$" (where these are the defined-up notions of existence), there remains a question of what exists in the distinguished sense, that is, *what really exists*.[15]

Hirsch, Reichenbach, and Kilo's claims of equivalence have been seen to turn on whether certain concepts (a unit, a metric, an existence-concept) are "distinguished". But if our goal is to clarify what we mean by equivalence, we haven't made much progress. For what does it mean to call a unit of measurement or a metric or an existence-concept 'distinguished'?

The idea itself is a familiar one. We don't normally think of space as having a distinguished direction or origin, whereas we (most of us anyway) *do* think of space as having a distinguished, or "intrinsic", metric. But what exactly does that amount to?

---

[14] Actually van Inwagen did accept some composites: living things.
[15] See Sider (2001, introduction; 2009; 2011, chapter 9).

To move forward, we'll need to enter some disputed territory in metaphysics. This shouldn't be a surprise. Investigating the idea of theories "saying the same thing about the world" leads immediately to general questions of what sameness and difference in the world consist in, very abstract questions about what the world's ultimate constituents are, about how we should think about such matters . . .—the deep seas of metaphysics. I doubt that a substantive metaphysics of equivalence can avoid broaching such issues.

5.3 Fundamentality

I think the best way forward makes use of the metaphysical tool of concept-fundamentality: equivalent theories are those that say the same thing about the world at the fundamental level, those that give the same description of the world in terms of fundamental concepts.[16] Thus theoretical content is individuated metaphysically, as upshots for fundamental reality. Further, to say that a unit, metric, or existence-concept is distinguished is to say that it is fundamental, or at any rate that it is uniquely singled out by a fundamental description of the world. A believer in a distinguished unit of measurement thinks, perhaps, that there is an absolutely fundamental relation between massive objects and real numbers, the mass-in-kilograms relation, and that relations like mass-in-grams and mass-in-pounds are not absolutely fundamental. Suppose an object is 1 kilogram in mass. Then the fact that it bears the mass-in-kilograms relation to the real number 1 is a fundamental fact. That object will also bear the mass-in-grams relation to the real number 1,000, but this fact is not fundamental; it holds in virtue of the fact that the object bears the mass-in-kilograms relation to 1.

An account of 'saying the same thing about reality' must specify the relevant sense of 'reality'. This we have done: the sense, according to the present approach, is that of fundamental reality. But it must also specify the relevant sense of 'saying', and this we have not yet done. 'Saying' had better signify some metaphysical relation, not ordinary saying, since the theories might not "say" anything at all about fundamental reality in the ordinary sense (section 5.1). Exactly which metaphysical relation will depend on our general approach to the connection between fundamental and nonfundamental. We might, for instance, speak in terms of ground, which for these purposes should be understood as *full, nonfactive, weak, mediate* ground in Fine's (2012, section 1.5) terms; propositions p_1 and p_2 can then be said to be equivalent when, for any propositions that involve only fundamental concepts, those propositions ground p_1 if and only if they ground p_2.[17]

[16] Miller (2005) defends a similar view.

[17] The grounding must be nonfactive to avoid 'Snow is white' being equivalent to 'either snow is white or grass is purple'. It must be weak to ensure that a statement involving a fundamental concept that lacks a strict ground can be equivalent to a statement that involves a nonfundamental concept which is defined in terms of the fundamental one.

An alternative approach would be to assume that any sentence has a "fundamental equivalent", a statement (or set of statements) using only fundamental concepts that gives the fundamental upshot

Another thing that must be sharpened is the claim that a distinguished concept is one that is "uniquely singled out" by the fundamental concepts. For example, we might want to speak of distinguished concepts of chemistry, because those concepts have simple definitions in terms of fundamental concepts. But 'simple definition' is itself unclear, and we may also wish to speak of distinguished concepts in domains in which simplicity of definition in fundamental concepts isn't the whole story.[18] For present purposes, though, it will be enough to work with a sufficient condition: a concept is distinguished if it is fundamental.

Equivalence, as understood by the fundamentality approach, is "nontransparent".[19] Suppose that mentality is in fact a matter of chemistry, which in turn is based in physics. Certain sentences about the mental might then count as equivalent to certain sentences about chemistry, because they "say" the same thing about (absolutely[20]) fundamental reality. Despite this, a rational person need not regard those sentences as having the same truth value. A dualist, for example, could assert one and deny the other without thereby being rationally deficient or conceptually confused. For the relevant sense of 'saying' the same thing about fundamental reality is metaphysical; what a theory says in this sense about fundamental reality is not something that we are guaranteed epistemic access to. Theories can therefore be epistemically or conceptually inequivalent in some perfectly good sense but nevertheless, given how the metaphysical chips have fallen, not correspond to any distinction in fundamental reality. For an account of equivalence in the philosophy of science, these features seem acceptable.

Relatedly, equivalence is "nonmethodological", given the fundamentality approach. On that approach, the claim that theories are equivalent rests on a substantive metaphysical assumption, namely that fundamental reality lacks the structure to privilege one of the theories. The claim that theories differing solely by the unit for measuring mass are equivalent presupposes, for instance, that Kilo's metaphysics is wrong. We might have wished instead for a conception of equivalence under which anyone adhering to broadly speaking scientific methodology would be guaranteed to agree on what is equivalent to what. One naturally or naively wants out of a notion of equivalence a weapon to *silence* anyone who wants to think about absurd questions such as whether a 1 kg object is really 1 or 1,000 in mass. But such a weapon—conceived as being effective regardless of target—cannot

---

of that sentence, the constitutive necessary and sufficient conditions at the fundamental level for that sentence. In terms of this notion, we could say that statements are equivalent when they have the same fundamental equivalent.

[18] Parallel issues are discussed in Sider (2011, section 7.11.1). See Gómez Sánchez (2021) for an account of this sort.

[19] Thanks to Chris Hauser and Michaela McSweeney here.

[20] I have been using 'fundamental' in the sense of absolute or perfect fundamentality. One might introduce some less demanding notion of fundamentality, on which concepts from the special sciences (Gómez Sánchez, 2021) or other "high-level" domains of discourse count as fundamental, and introduce a corresponding notion of equivalence. One could then recognize more cases of inequivalence (particularly weak inequivalence—see below) than with my official notion of equivalence.

exist. Nothing one could say about the nature of equivalence can simply silence Kilo, since Kilo has an internally coherent conception of the nature of sameness and difference of "metaphysical content" which allows differences in content corresponding to units of measure. This is not to deny that Kilo is irrational in some sense; his metaphysical outlook, after all, is surely not supported by the evidence. Nevertheless the rejection of his point of view is a substantive, metaphysical stance, not the result of mere reflection on methodology.

5.4 Difficult choices

This account of equivalence in terms of fundamentality is the first of the two "extreme" approaches I want to discuss. I call it extreme because it has certain uncomfortable consequences.

First consider an example in which all goes well. There is surely no distinguished unit of measurement for mass; theories employing different units are equivalent (provided their numerical values are appropriately related). Given the fundamentality approach, those theories must say the same thing at the fundamental level; they must express the same facts, stated in terms of fundamental concepts. What are those fundamental concepts, the fundamental concepts of mass? Various plausible answers are available, with help from the theory of measurement, as we saw in Chapter 4. For instance, one could hold that $x \succeq y$ (x is at least as massive as y) and $Cxyz$ (x and y's combined masses equal z's) are the only fundamental concepts of mass. Uniqueness theorems assure us that numerical theories of mass based on different units code up the very same $\succeq$- and C-involving facts.

But notice a feature of the example. To support a claim of equivalence between a pair of theories, stated in a pair of languages in which mass is described using different units, we brought in a *third language*, a language in which mass is described in a unit-free way, using the concepts $\succeq$ and C. This third, more[21] fundamental, language gave us a perspective on the fundamental facts, a perspective from which the first two theories could be seen as getting at the very same facts. But what if there is no such a third language? For instance, what if one or both of the theories we are considering for equivalence are already in perfectly fundamental languages?

In some such cases, the "third" language could be the language of one of the two equivalent theories, if that language can underwrite the language of the other theory. Any theory in a fundamental language L is equivalent to some "gruefied" theory T';[22] the "third" language here is just L. But what if both languages are fundamental?

---

[21] The third language needn't in general be *perfectly* fundamental. We might have reason to believe that the fundamental grounds for the claims of two theories "run through" some more-but-not-perfectly-fundamental language.

[22] Compare the equivalence of 'all emeralds are green' and 'all emeralds are either grue and first observed before the year 3000 or bleen and not first observed before the year 3000'.

For instance, consider the case of ontology. Hirsch wants to say that the theories of Lewis and van Inwagen are equivalent—that although they appear to be incompatible theories stated in a common language, they are in fact equivalent theories stated in distinct languages—languages in which the quantifiers mean different things. But the dispute between Lewis and van Inwagen is a dispute over *what there is*, which is apparently as fundamental a concept as can be. So it is hard to see how there could be a third language from whose perspective Lewis and van Inwagen can be seen as getting at the same facts. If a proposed "third" language is either Lewis's or van Inwagen's language, or indeed any quantificational language, then the facts stated in that language will fail to vindicate either van Inwagen or Lewis (or both), and it won't be the case that each is getting at the same such facts.[23] The third language must somehow be "pre-quantificational", so as to be "neutral" between van Inwagen and Lewis, in the way that the comparativist language of $\succeq$ and C is "pre-numerical" and neutral between numerical theories based on different units. It must be that Lewis's and van Inwagen's quantificational languages are nonfundamental, and are underwritten by some more fundamental language that lacks quantifiers (or anything like them). In light of Chapter 3, this conception of fundamental reality is unlikely to be true.[24]

---

[23] In more detail: let Lewis's claim be (L): 'every two objects are part of some object', let van Inwagen's claim be (I): 'it's not the case that every two objects are part of some object', call Lewis's language L and van Inwagen's language I, let $\langle A \rangle_I$ be the proposition expressed by sentence A in language I, and suppose for reductio that Lewis and van Inwagen's claims are equivalent—that is, that $\langle (L) \rangle_L$ is equivalent to $\langle (I) \rangle_I$. Given the ground-theoretic version of the fundamentality account of equivalence given in section 5.3, it follows that (*) for any proposition, p, that can be stated in the third language, p is a full, nonfactive, weak, mediate ground (henceforth simply 'Ground') of $\langle (L) \rangle_L$ if and only if p is a Ground of $\langle (I) \rangle_I$. We are assuming that the third language is quantificational, by which I mean that it contains the standard quantifiers. For simplicity, I'll also assume it contains the predicate 'is part of'. (This is dispensable; see Sider (2009, p. 390).) Thus the third language contains the very sentences (L) and (I). Since (I) is the negation of (L), one or the other must be true in that language; suppose without loss of generality that it is (L). Now, both van Inwagen and Lewis mean to be disagreeing about "what there ultimately is". So it is implicit in their usage of quantifiers (and anyway they would be willing to stipulate this) that their quantifiers are intended to be "ontologically perspicuous" (see Sider (2009) for issues in this vicinity). Here we may take this to require that their quantified sentences must be "homophonically grounded": the proposition expressed by a quantificational sentence in either of their languages must be grounded by the proposition expressed by that very sentence in any quantificational fundamental language, if there is such a fundamental language. Thus, if we call the third language '3', we have that $\langle (L) \rangle_3$ Grounds $\langle (L) \rangle_L$, and that $\langle (L) \rangle_3$ Grounds $\langle (L) \rangle_I$. By the latter and the fact that (L) is true in 3, (L) is true in I. By the former and (*), $\langle (L) \rangle_3$ Grounds $\langle (I) \rangle_I$, and so (I) is true in I. But (I) is the negation of (L). Assuming that no contradictions are true in I, the reductio is complete.

[24] The most plausible entity-free views considered in Chapter 3, bare particulars and generalism, are of no use here since they aren't really "pre-ontological" in the relevant sense; we can raise the questions of composition using vocabulary that those views deem fundamental. For instance, if parthood is a fundamental relation, then we can translate 'Every two objects are part of some object' into Dasgupta's term functorese. The resulting sentence would be a claim about fundamental reality, and would be either true or false; and either way, the argument can proceed as in note 23. The facts statable in "quasi-quantificational" languages like term functorese are not "neutral" between Lewis and van Inwagen.

The fundamentality approach, then, arguably rules out Hirsch's view that Lewis and van Inwagen's ontological claims are equivalent. Now, there is a long discussion to be had here, but my own view is that this is the correct verdict. Lewis and van Inwagen's ontological claims are inequivalent precisely because there is no neutral, "pre-ontological", fundamental theory from whose point of view those claims can be seen as getting at the same facts. Competing ontological claims are inequivalent; there is a genuine issue about whether composite objects really exist.

More generally, the fundamentality approach implies that when apparently conflicting claims are cast in terms of perfectly fundamental concepts, the conflict is genuine, and cannot be resolved by a Hirschian claim of equivalence. (Of course, one can be mistaken about whether one's concepts are fundamental.) Some will reject this implication of the fundamentality approach, but I don't myself think there is serious dialectical pressure here; at least, not for those who are open to the notion of fundamentality itself.

But in other cases there is more pressure. Like the case of ontology, they involve theories stated in languages that at least appear to be perfectly fundamental. But unlike the case of ontology, the theories are not in apparent conflict (a conflict to be resolved by a Hirschian claim of equivalence). They are, rather, theories stated in distinct vocabularies; and practically everyone believes them to be equivalent.

For example, let $T_\forall$ be some theory in which the only quantifier is $\forall$, and let $T_\exists$ be the result of replacing each occurrence of $\forall v$ in $T_\forall$ with $\sim\exists v\sim$. This is a paradigm case of equivalence, one might naturally assume. But it might seem that the fundamentality approach cannot allow this. For the languages in which these theories are stated, $L_\forall$ and $L_\exists$, are quantificational, and as before there is reason to doubt the existence of a more fundamental pre-quantificational third language underlying each.

Now, this challenge can be answered in its current form. As we noted earlier, the "third" language can in fact be one of the languages of the theories in question. For instance, suppose universal quantification is a fundamental concept and existential quantification is not. Then $L_\forall$ could be the "third" language. The facts statable in $L_\exists$ would be underwritten by the facts statable in $L_\forall$; $\exists vA$ would be (strictly) grounded in the more fundamental $\sim\forall v\sim A$. Thus the fundamentality approach can allow the equivalence of $T_\forall$ and $T_\exists$ after all. Each of them gets at the same fundamental facts, namely the facts statable in $L_\forall$.

But the reprieve is only temporary. Let us distinguish between two sorts of equivalence, weak and strong. Weak equivalence is the kind of equivalence that we have been discussing so far. Strongly equivalent theories are those that, in addition to being weakly equivalent—in addition to saying the same thing at the fundamental level—do so "equally perspicuously". I won't attempt a rigorous definition, but the rough idea is that the perspicuity of a theory is a function of the fundamentality of the primitive concepts of the theory's language. Thus even though $T_\forall$ and $T_\exists$ are weakly equivalent, they are not strongly equivalent: $T_\forall$ is more perspicuous than $T_{\exists}$ since $L_{\forall}$'s primitive quantifier is perfectly fundamental and $L_{\exists}$'s is not (and since $L_{\exists}$ is otherwise exactly the same as $L_{\forall}$). And admitting *any* sense in which these theories are inequivalent is difficult to accept.[25]

Really, though, there is a more primordial concern here, which is not that the fundamentality approach to equivalence makes the wrong predictions, but rather that its conceptual apparatus generates problematic questions. If we are willing to speak of the fundamentality of concepts such as universal and existential quantification, then we must face the question of *which* is fundamental; and this question will seem to many to be absurd. How could there be facts such as that universal quantification is a fundamental concept but existential quantification is not?

An advocate of the fundamentality approach can always reply that there are facts here of which we are ignorant. One or the other of existential and universal quantification, or perhaps a third concept of some novel sort, as Michaela Mc-Sweeney (2019) urges, is the fundamental concept in the vicinity, but we simply do not know which concept that is, nor need we say why it is the fundamental one. It is natural to claim that fundamentality is itself fundamental, in which case no grounds should ever be demanded for the claim that a concept is fundamental; and although there usually is evidence available for claims about fundamentality, there is no reason to suppose that such evidence is always available. But these standard "realist" lines may seem particularly hard in the present case. After all, the choice between which quantifier to take as basic—and similarly for the propositional connectives—is usually regarded as a paradigmatically conventional one. You

[25] We should revisit the case of Kilo in light of this distinction. I said earlier that Kilo thinks that statements about grams are inequivalent to statements about kilograms. But Kilo presumably thinks that these statements get at the same facts about the distinguished unit; so at best they fail to be strongly equivalent. However, the case is still more complicated. Kilo's statement '*o* is 1 in mass' should be distinguished from the statement '*o* is 1 kg in mass', in which the kilogram unit is mentioned explicitly. Presumably Kilo's view would be that when ordinary people—who do not believe in a distinguished unit—say things like '*o* is 1 kg' and '*o* is 1,000 g', then these statements are in fact strongly equivalent, on the grounds that neither is more perspicuous than the other since neither is about the distinguished unit. (He would take the same attitude towards this pair of statements that realists about the metric would take towards a pair of apparently conflicting statements about distance that explicitly build in Reichenbachian coordinative definitions.) To get a pair of statements that Kilo would regard as inequivalent, we should consider statements made by people who believe in a distinguished unit but disagree over what that unit is. Imagine Kilo says '*o* is 1 in mass', and his friend Gram, who thinks the gram is the distinguished unit, says '*o* is 1,000 in mass'. But note that if Kilo's claim that he and Gram are making inequivalent statements is to have the proper force—as a claim that is distinctive of those who believe in a distinguished unit—then Kilo and Gram's statements can't be understood as being too metaphysically loaded, as explicitly speaking of a distinguished unit. For even those who reject a distinguished unit of mass can agree that statements that explicitly say that there *is* a distinguished unit, and go on to make competing claims about what amount of mass in that unit is had by *o*, are inequivalent. Kilo and Gram's attitude towards their numerical statements about mass must be those of scientists, not metaphysicians: they view numerical talk about mass in the same spirit in which most scientists normally view talk of spatial congruence, needing neither coordinative definitions nor explicit mention of fundamentality.

The example of Kilo is more subtle than it first appears!

don't need to be a logical positivist to feel that metaphysics has gone off the rails if it leads to questions like whether universal or existential quantification is more fundamental. And it is similarly hard to hold out hope for some third concept underlying those two.[26]

In some cases there may be theoretical considerations favouring one rather than another choice, which can be regarded by the friend of fundamentality as super-empirical virtues of claiming that the chosen concept is more fundamental. But it is hard to believe that there will always be some such considerations.

The epistemology of fundamental concepts from section 1.8 is no help here. It tells us to, for instance, regard concepts as fundamental when they are needed to formulate simple and strong laws. But rarely if ever does this need single out a single concept. The situation, rather, is that some concept or other of a certain sort is needed. In logic we need something like the standard quantifiers, but either the existential or the universal quantifier will do, so the epistemology tells us we must adopt one or the other (or some other concept that can do similar work (Donaldson, 2015)) without telling us which.

The problem is ubiquitous. In addition to the standard universal and existential quantifiers, there are other conceptual bases for quantification theory, such as the predicate and term functors discussed in Chapter 3, and quantifiers that do not bind variables but rather are applied directly to predicates formed by lambda abstraction (Stalnaker, 1977). There are multiple bases for propositional logic. And the problem isn't restricted to logic. Instead of $\succeq$ and C, a theory of mass could be based on any of their converses. This instance of the problem can perhaps be solved by the view that a relation is identical to any of its converses (Williamson, 1985; Fine, 2000; Dorr, 2004), but other instances of the problem remain. For example, if reality has fundamental mereological structure, we face the question of which mereological concept is fundamental, parthood, overlap, or fusion; and no two of these are converses. In such cases, many will want to say, the question of which conceptual decision to make is conventional, and theories based on alternate conceptual decisions are in every reasonable sense equivalent.

5.5 Quotienting

What, though, is the alternative to the fundamentality-based approach? According to one alternative—the second "extreme" approach to equivalence I want to discuss—we can say *that* theories are equivalent without saying *why* they are equivalent in terms of fundamentality and underlying third languages.[27] What *is* equivalence, according to this view? No illuminating account can be given,

---

[26] See Donaldson (2015); Dorr and Hawthorne (2013, section 4); McSweeney (2016; 2019); Sider (2011, section 10.2); Steward (2016); Sud (2018); Torza (2017); Warren (2016) on this and related issues.

[27] Dewar (2015; 2019b) defends a view closely related to quotienting. The latter paper squarely engages the central issue I'm concerned with here: whether it's appropriate to demand a "third, more fundamental language", or in his terms, a "reduced" theory.

according to this approach (which is not to say that equivalence is somehow a metaphysically fundamental notion).

Let's consider an example. In the case of ∀ and ∃, the defender of this second approach might make the following speech:

A good theory can be formulated using the concept of ∀. But one can formulate an equivalent theory using the concept of ∃ instead. Indeed, we can define a relation between theories that guarantees equivalence: differing solely by exchanges of formulas QvA and $\sim Q'v \sim A$ (for Q one of ∃ and ∀, and Q' the other). True, we cannot provide a third, "more fundamental" description of quantificational reality underlying this relation. But no such description is needed; it's enough simply to say that theories standing in the relation are equivalent.

Similarly, one could say that any theories differing solely by a unit of measurement are equivalent, or that Lewis and van Inwagen's theories are equivalent, without saying *why* these theories are equivalent via a third, more fundamental language, and even without believing that there is any such language. 'But what are you saying reality is like?', you might protest. The answer will be that reality is such as to be well described in any one of the equivalent ways, and that there is no need to say anything further.

I think of this view of equivalence as flowing from a certain view about the nature of modelling, realism, and so on, concerning the way one separates representational content from "artefacts of the model". Everyone agrees that a good model can have features that aren't part of its representational content. A map of the USA drawn on paper doesn't represent the USA as being made of paper. Or suppose one chooses to represent mass using kilograms. One is then using real numbers as a kind of model, a model in which, for example, the number 1 represents the mass of certain things (1 kg things). But the fact that the number 1 is used isn't part of the representational content of the model; it's an artefact of the choice to use one scale rather than another for measuring mass. The objects aren't objectively 1 in mass, assuming there is no distinguished unit.

I think that many metaphysicians tend to assume (perhaps implicitly) something like the following:

It's fine to construct models with artefacts. But there must always be some way of describing the phenomenon in question that (in some sense[28]) lacks artefacts. There must be some way of saying what is really going on. For example, although we can model mass with real numbers, there must be some underlying artefact-free description, such as the $\succeq$ and C description, from which one can recover a specification of which numerical models are acceptable, and a specification of which features of the models are artefacts.

Whereas the second approach to equivalence I am describing here rejects this assumption, and says instead:

---

[28] In some sense *every* representation has artefacts—ink color of written sentences, tonal contours of spoken ones, and so forth. A first pass at the relevant sense of being artefact-free might be that the only artefacts are semantically inert. (Sider (2011, pp. 221–2) wrestles with a similar issue.)

There may be no way to say what is "really" going on, since in some cases, every good model has artefacts. It's then OK to just say: this model does a good job of representing the phenomenon, but certain features of the model are artefacts. Moreover, for any model, we can say which features of the model are genuinely representational and which are artefacts. There is no need to provide some privileged, artefact-free description from which we can recover this information.

(Some of this is reminiscent of the semantic view of theories, but it seems to me that, whether theories are sentence-like or model-like, there will arise the issue of whether a demand for a certain sort of best theory is legitimate.)

Think of it this way. If we have multiple theories with conventional differences, the advocate of this second approach says that one can "quotient out" the conventional content, and replace the theories with their equivalence class.[29] Putting forward the class, rather than any one of the theories, is representationally superior, for one thereby fails to commit to any particular conventional choice. Moreover, one can quotient out the conventional content "by hand": the equivalence relation doesn't have to be induced by some more fundamental theory—as I think it should be—but rather can simply be stipulated. When faced with equivalent formulations of theories, instead of seeking theories that explain why the formulations are equivalent, the quotienter is satisfied with equivalence classes of theories, in effect passing from the original set of available theories to its quotient set in the mathematical sense.

The quotienter I primarily have in mind rejects all talk of fundamentality, and as a result has an expansive conception of equivalence. Recall Goodman's (1955b) grue/bleen example:

> An object is *grue* if and only if it is green and first observed before 3000 AD or blue and not first observed before 3000 AD.
>
> An object is *bleen* if and only if it is blue and first observed before 3000 AD or green and not first observed before 3000 AD.

Goodman points out that although these equivalences define grue and bleen in terms of green and blue, one could reverse the definitions:

> An object is *green* if and only if it is grue and first observed before 3000 AD or bleen and not first observed before 3000 AD.
>
> An object is *blue* if and only if it is bleen and first observed before 3000 AD or grue and not first observed before 3000 AD.

---

[29] What I am calling quotienting should be distinguished from another activity that sometimes goes by that same name, namely the construction of entities as equivalence classes of other entities, such as directions as sets of parallel lines, or rational numbers as sets of ordered pairs of integers under the relation $\langle m, n \rangle \sim \langle o, p \rangle$ if and only if $m\,p = o\,n$. When one constructs entities in this way, one is (normally) putting forward a single theory of what the entities in question are (namely that they are sets), whereas the quotienter in my sense puts forward a class of theories and claims they all represent the world equally well. Thanks to Isaac Wilhelm here.

In addition to saying that a theory stated in terms of ∀ is equivalent to one stated in terms of ∃, and that a theory stated in terms of parthood is equivalent to one stated in terms of overlap, the quotienter I primarily have in mind would also say that a theory stated in terms of grue and bleen is equivalent to one stated in terms of blue and green, or even that theories with apparently conflicting ontologies can be equivalent, and would reject the idea that any of these theories are "more fundamental" (or "carve the world at its joints" better) than any of the others. This quotienter utterly rejects "the ready-made world".[30]

But a milder quotienter might try to hang on to some talk of fundamentality, and some part of the ready-made world, and as a result have a narrower conception of equivalence. For one might admit that an account of fundamental reality ought to speak of green and blue rather than grue and bleen—or better, of quantities from physics rather than their "grueified" versions—while being unhappy with the fine-grained distinctions my own view leads to, and reach for quotienting as an intermediate position. One might, for instance, allow a distinction between fundamental and nonfundamental facts (facts about physical quantities are fundamental, facts about gruified versions are not) but say that some fundamental facts are equivalent to others ('some things have mass' is equivalent to 'not all things lack mass') while rejecting the demand for a "third language", a single, privileged description of fundamental reality. I myself think that this milder view is dialectically unstable, but it nevertheless should remain on the table.[31]

In many cases a quotienter can agree with the modal approach to equivalence (section 5.1). For instance, the modal approach implies that $T_\forall$ is equivalent to $T_\exists$, and the quotienter is free to agree. Nevertheless the approaches are distinct. First, the quotienter does not *need* to say that any theories that are true in the same possible worlds are equivalent. A quotienter who is a theist might, for instance, think that 'God exists' and '$2+2=4$' are inequivalent despite regarding them as modally equivalent. (For that matter, nothing in my presentation of quotienting *requires* a quotienter to say that $T_\forall$ is equivalent to $T_\exists$.) Further, it is available to a quotienter to regard some modally inequivalent statements as nevertheless being equivalent. For instance, a quotienter might accept a "mereological nihilist" possible world just like the actual world subatomically but lacking any composite objects like chairs, but nevertheless say that actual reality is equally well represented by 'there are subatomic particles arranged chairwise' and 'there are subatomic particles arranged chairwise and there is a chair'. According to such a quotienter, modality is sensitive to merely conventional differences. To

---

[30] This is in the spirit of Goodman (1955a; 1978) and Putnam (e.g., 1981) of course.

[31] The milder quotienting view should be distinguished from the acceptance of redundant fundamental structure. One might react to the problems of section 5.4 by, e.g., accepting *both* ∀ and ∃ as fundamental concepts. (See section 5.8.4 below and Sider (2011, section 10.2).) On this view, statements in terms of ∀ would of course be *logically* equivalent to their duals in terms of ∃, but they wouldn't be strongly equivalent in the sense at issue in this chapter, since they involve distinct fundamental concepts.

be sure, some quotienters will deny this, and regard modal equivalence as being necessary for equivalence. (Other necessary conditions might be proposed, such as having the same truth value and being a symmetry of the laws.) But this necessary condition isn't built into quotienting. Finally, a defender of the modal approach might admit that modal equivalence requires, in some or all cases, a postmodal explanation, whereas I am construing the quotienter as rejecting such explanatory demands.

5.6 The significance of quotienting

In my view the question of whether quotienting is legitimate has profound implications for a wide range of questions in the metaphysics of science, and within metaphysics generally. As I see it, the question is a crucial choice-point for foundational theorizing about science, which has generally lurked below the surface, but in retrospect can be seen to underlie various disputes, especially when some disputants are more hostile to metaphysical inquiry than others and there is a feeling that the disputants are talking past one another. Speaking just for myself, once I got the issue of quotienting clearly in view, a number of otherwise extremely perplexing disputes were dramatically clarified. I feel as though I finally understand what is going on.

For a very simple example of this, consider statements of the form 'only such-and-such an equivalence class is real' or 'what is real is that which is common to all members of the equivalence class'.[32] Language like this is quite common (especially in the philosophy of physics), but is perplexing. The first sentence is absurd if taken literally, and the second uses a noun phrase ('that which is common . . .') without a clear referent. But they make perfect sense if taken as expressions of quotienting.

In the following sections we'll examine some more substantive illustrations of the importance of the question of quotienting. But many other examples could be given: the interpretation of gauge theories, the interpretation of various dualities, and the status of intuitively extraneous mathematical objects in characterizing geometric structure, to name just three.[33]

5.6.1 *Quotienting and quantum mechanics*

One illustration comes from the dispute over the metaphysics of the wave function in quantum mechanics. The most straightforward metaphysics of the wave function treats it as a fundamental field, living in a fundamental, substantival high-dimensional space (Albert, 1996; 2015). But since this high-dimensional space is

---

[32] For instance, Saunders (2003, p. 154) describing Stachel's (1993) response to the hole argument: 'we should pass to the equivalence class of solutions under diffeomorphisms, a view which is by now quite standard in the literature . . .. Only the equivalence class is physically real.'

[33] See, for example, Healey (2007) on gauge theories, and Arntzenius and Dorr (2011) and Maudlin (2014) on geometry.

wholly distinct from the three-dimensional space of ordinary experience, many regard this straightforward metaphysics as unbelievable.

One might instead regard the wave function as being a field in abstract configuration space, a constructed space whose "points" are set theoretic constructions that represent the locations of all particles in three-dimensional space. But this approach is unattractive on its face, and rarely pursued. Perhaps many have bypassed it because it attaches a fundamental physical magnitude to abstract entities. I myself think its main vice stems rather from the arbitrariness of the construction of abstract configuration space: on which of the many abstract entities that can with equal justice be regarded as configuration space is the wave function defined?[34]

But for a quotienter, the metaphysical status of the wave function is not at all puzzling. Indeed, the quotienter will likely regard angst over the status of the wave function and the space in which it "lives" as deriving from an illegitimate demand for metaphysics. For one can simply say the following:[35]

You got into trouble because you were trying to specify, once and for all, what the fundamental properties and laws were, independently of conventional choices. That's misguided. The proper procedure is instead the following. First we must make some arbitrary choices for how to construct configuration space. Given those choices, there is a certain physically significant wave function on configuration space. But we cannot, and need not, say what the physical facts and laws are, prior to making conventional choices.

More fully: there are various theories, $T_1, T_2, \ldots$ based on different equally acceptable constructions of abstract configuration space. In each case, the theory T_i will utilize a physical predicate Ψ_i for wave function values at points in abstract configuration space as constructed in T_i. And although a relation of equivalence can be defined between the theories, which will constrain the relationships between the physical predicates Ψ_i, there is no definition of any of the predicates Ψ_i in terms that don't refer to the others. This may all seem unsatisfying. 'Which aspect of physical reality are these predicates Ψ_i representing?', you ask. Surely physics is about how the world is, in and of itself? But I am not denying this. The world is such as to be representable by a wave function in an abstract space, by any of the theories T_i; and this is the case independent of any conventional choices. But there is nothing further that one can say, nor is there any need to. There is no "God's-eye" description of reality.

Thus the quotienter would be denying the need to ground the use of mathematical representations of the wave function with a representation theorem; we can stop with the mathematical representations.

Indeed, one might take this a step further, and regard even the question of whether the "fundamental space" is ordinary three-dimensional space or some high-dimensional space as itself being conventional. On this view, there's nothing more to say about the "real" dimensionality of physical reality, and about the ontology of space, other than that reality can be described as containing a concrete

---

[34] For a discussion of parallel issues see Sider (1996).

[35] The quotienting outlook seems to be best conveyed through speeches.

configuration space with a complex-valued field on it, but can also be described as containing a concrete three-dimensional space, in which the wave function is modelled in some more complicated way.[36]

David Wallace in fact takes something like this line. He writes of 'a gap in the market for some intermediate philosophical position, one which respects scepticism about overly "metaphysical" claims while incorporating the impossibility of any coherent theory/observation divide' (Wallace, 2012, p. 314). What he has in mind is the following. On one hand, he thinks that the question of whether space is *really* three-dimensional is not a genuine one (it's "overly metaphysical"). Similarly, many would say, the question of whether Lewis is right that there exist tables and chairs, or whether van Inwagen is right that there do not, is not a genuine question. But on the other hand, the most familiar way of rejecting there being a genuine question here is that of the discredited positivists, whose main sin, according to Wallace, is upholding a sharp divide between theory and observation.

Wallace goes on to say that he suspects the gap in the market will be filled by structural realism, but I don't think that's right. I read the structural realist literature as mostly accepting a more standard "metametaphysics"—accepting a demand for a (mostly) artefact-free metaphysics. Otherwise why take so seriously worries about the coherence of a conception of metaphysics without objects, relations without relata, and so forth? What Wallace is really after is a much more conventionalist attitude towards ontology and metaphysics in general; quotienting would provide that.

5.6.2 *Quotienting and ontology: Hirsch*

According to Hirsch's quantifier variance, apparently incompatible ontological claims can be equivalent if the quantifiers in those claims have different meanings. Hirsch would say, for example, that 'there exist chairs', given what Lewis means by 'there exists', is equivalent to 'there exist subatomic particles in a certain arrangement C', given what van Inwagen means by 'there exists'. As I said above, my main concern about this approach is that there seems to be no "ontologically neutral" third language available to Hirsch to state what it is about fundamental reality that van Inwagen and Lewis both describe using their sentences.

Hirsch sometimes gestures at a conception of "unstructured facts", which might seem to be the basis of a suitable third language: '[W]e can retain the notion of an unstructured fact. I think this is indeed our most basic notion of "reality", "the world", "the way it is", and this notion can remain invariant through any changes in our concept of "the things that exist"' (2002, p. 59). However, he never develops a metaphysics of unstructured facts in any detail. (Which may well not be inappropriate; I suspect he did not intend in such remarks to advocate for a fundamental metaphysics of unstructured facts.) And it is hard to see how

---

[36] Note the conventionalism about ontology. This does not require antirealism in the sense that reality is up to us; see Hirsch (2002).

such a metaphysics *could* be suitably developed. The conception of facts would need to be both foundationally adequate and also "neutral" with respect to the Lewis/van Inwagen dispute, and thus would need to be "pre-ontological" in the sense discussed earlier.[37]

I don't see how Hirsch could meet this objection, if he accepted the fundamentality-based approach to equivalence. But he could easily meet it if he were a quotienter. He could then simply claim that theories in various ontological languages are equivalent, without needing to provide a more fundamental conception of the underlying facts. Quotienting would seem to fit Hirsch's general metametaphysical outlook, and could even be seen as underlying his remarks about unstructured facts.

5.6.3   *Quotienting and modality: Stalnaker*

Another nascent instance of the quotienting strategy, I believe, can be found in Robert Stalnaker's book *Mere Possibilities*.[38] Stalnaker accepts a possible worlds semantics for modal concepts like possibility and necessity. Moreover he is an "actualist": he rejects Lewis's (1986a) modal realism and holds that possible worlds are just parts of actuality: they are properties, he says—ways reality could be. Now, there is an old problem with this approach: how to construct possible worlds containing entities that do not in fact exist? Consider possibilities in which two nonactual dice—dice that are distinct from every actual entity—are rolled so that their sum is 3. There should be two possible worlds here: one where one die comes up 1 and the other comes up 2, and a second world where the first die comes up 2 and the second die comes up 1. But since these dice do not in fact exist, it is hard to see how there really could be two worlds, if worlds are just properties and actualism is true, since there do not seem to be materials in the actual world to construct distinct properties to identify with the worlds.

Stalnaker's solution to this problem is intriguing. He accepts the usual approach to possible worlds semantics, in which formulas are evaluated for truth value relative to the members of a certain set, W. But instead of calling those members 'worlds', Stalnaker instead calls them 'points'. The set of points corresponds, intuitively, to what a *non*-actualist would regard as all the possible worlds. In particular, even though we cannot construct distinct world-properties for the dice summing to 3, Stalnaker recognizes two points here: one for 2/1, the other for 1/2. What *are* these points? The particular entities that are the points, Stalnaker says, are unimportant—their identity has no particular representational significance in the model. They could be bananas, or fish, or numbers; "it's only a model". Moreover, neither point individually represents anything different from what the other point represents. Apart from the rest of the model, each represents the

---

[37] See Sider (2011, section 9.6.2) for more on all this. The required neutrality means, for example, that the fascinating theory in Turner (2016b) would not suit Hirsch's needs.

[38] In fact, my conception of quotienting crystallized while thinking about this book.

same thing, namely the possibility of there existing two duplicate nonactual dice summing to 3. For there simply *are* no possibilities of *particular* nonactual dice *A* and *B* coming up 2/1, say, for them to differentially represent, since there do not in fact exist dice *A* and *B*. But the fact that the model contains *two* points here rather than one *is* representationally significant: the model as a whole represents the modal fact that had two such dice existed, there would have been two possibilities.

Since the points corresponding to nonactual possibilities (so to speak) are not representationally significant, there are different Stalnakerian models that are representationally equivalent. Stalnaker therefore provides an account of which of his models are representationally equivalent to others—thus, I would say, quotienting out the artefactual content. And he does this "by hand", in that he does not give any further account of what possible worlds are, or what properties are, or what the modal facts are, that renders this class of models apt or justifies his claims about which models are equivalent. His attitude is: modal reality is such as to be well modelled in this way, and there's no need to give any further, artefact-free account of modal reality that shows *why* this is the case.

Stalnaker is explict that a distinctive view about the nature of metaphysics, and of the division between substantive and conventional/semantic questions, is central to his approach.[39] Indeed, he compares his equivalence classes of possible-worlds models (differing by which entities play the role of which "mere points") to equivalence classes of models of space that would be given by a relationalist: the relationalist will regard models differing only by translations, for example, as being representationally equivalent. I think what is going on is that Stalnaker is a quotienter.[40]

5.6.4 *Quotienting and structuralism*

A final illustration of the importance of quotienting is the main topic of this book: structuralism. As we've seen in previous chapters, structuralist theses can be difficult to articulate in postmodal terms. Sometimes there simply is no attractive view of the nature of fundamental reality that is structuralist in spirit. But a structuralist who accepts quotienting could dismiss all those concerns as arising from an illegitimate demand for metaphysics. The structuralist slogan that nodes are not 'independent of' patterns could naturally be taken as a claim of equivalence: variation of nodes while leaving the pattern intact results in an equivalent theory. And given quotienting, instead of attempting to specify a structuralist-inspired conception of the fundamental facts from which these equivalences could be

---

[39] See, for instance, Stalnaker (2012, p. x).

[40] Consider also his resistance to there being a once-and-for-all answer to the question of how fine-grained propositions are (Stalnaker, 2012, pp. 12–13). This suggests the view that the modal facts may be represented by many models of varying granularity, but that there is no once-and-for-all conception of the facts underlying the models' adequacy. This is akin to quotienting but with this difference: the models are not all on a par; rather, they increase in accuracy in a certain sense as the grain becomes finer.

derived, one could simply define, by hand, an equivalence relation on descriptions of patterns. Many of the problems I raised in previous chapters would immediately be avoided, just like that.

In the case of quantity, for example, the quotienter could say that descriptions of the world's mosaic that differ by a mere choice of unit are equivalent, without saying why that's so.[41] It would be natural to also adopt "nomic quotienting" (section 4.9.3), and say that the laws are to be stated in the usual, textbook, numerical way, using any chosen units, without giving any account (in terms of more fundamental non-numerical laws) of why the laws have the same form regardless of the chosen units, or any account of the variation in the laws' constants when the units are varied. The concerns of Chapter 4 were about how to avoid such vices as arbitrariness, artificiality, nonlocality, and so forth in the laws. But the quotienter could respond that those concerns are an artefact of a misguided meta-metaphysics. They are about "spandrel questions", as Jessica Wilson (2018) put it in another context: questions stemming from the adoption of illegitimate vocabulary (such as that of fundamentality) which simply do not arise once that vocabulary is abandoned. The concerns arose from an attempt to find laws that are simple, local, nonarbitrary, nonartificial, and so on *when formulated in terms of fundamental concepts*, since such laws would need to involve quantification over standard sequences, or arbitrary choices amongst relations to numbers, or the like. Once we abandon the metaphysical tool of concept-fundamentality and no longer seek attractive laws formulated in terms of fundamental concepts, the concerns evaporate. The only bias in favour of simple laws that it is legitimate to insist upon, the quotienter will say, is a bias towards simple numerical statements of the sort familiar in textbooks. The only sort of locality we should seek, the quotienter will say, is of a straightforward numerical sort—in the case of Newton's dynamical law, for example, the determination of the numerical value of acceleration at an instant as a function of mass and the numerical value of force at that instant. And so on.

The impact of quotienting on my critique of nomic essentialism would be just as dramatic. A nomic essentialist could define a relation between theories that holds when those theories differ only by a permutation of scientific properties in the laws, and claim that theories standing in this relation are equivalent. She could stop with this kind of claim, rather than trying to develop a distinctively nomic essentialist conception of the fundamental facts. All my concerns about that latter project would thereby be avoided.

Likewise for the case of structuralism about individuals, in which this approach may be especially appealing. Instead of struggling to find an entity-free funda-

---

[41] If one thought that further differences are also merely conventional—for example doubling the mass of each thing, which is regarded by comparativists as a distinction without a difference—one could make further claims of this sort, again without the need for a distinctively comparativist conception of the fundamental facts, although this would require the enhanced sort of quotienting to be discussed in the next section.

mental account of reality, structural realists (for example) could simply claim that certain descriptions are equivalent—descriptions that differ solely by a permutation of individuals, for instance.[42] (Quotienting may be particularly welcome for structural realists like French (2014), whose conception of structure is highly "abstract".)

Relatedly, recall the discussion of "scooping out" structure in section 3.15. A conception of the physical structure of reality might have an "aspect" that the laws don't care about. The laws don't care about the identities of points of space or particles, for instance; structure-preserving permutations are symmetries of the laws. Nevertheless, given realism about fundamental concepts, this aspect can't simply be "scooped out" leaving the rest of the theory intact. We need to see whether the aspect corresponds to some fundamental concept that can be eliminated while keeping the theory attractive. But given quotienting, arbitrary aspects *can* be scooped out in this way. For one can simply define up an equivalence relation corresponding to the aspect in question.

As it happens, structural realists seem not to have availed themselves of the quotienting option. They have instead engaged in a more traditional metaphysical project, seeking an account of fundamental reality that eliminates the sorts of differences that they regard as nongenuine. The ontic structural realist slogan, after all, is that 'all that there is, is structure'.[43] (Quotienters embrace nonstructural features of reality; they just insist that there are many equally good ways to speak of them.) But I wonder whether the more deeply antimetaphysical approach of quotienting might be a better fit for their overall outlook.

5.7 Actual and counterfactual equivalence

The last two applications of quotienting from the previous section—to nomic essentialism and to structuralism about individuals—are not as straightforward as they might first seem. In each case I imagined the quotienter saying that descriptions that differ by a mere permutation are equivalent. But such descriptions often differ in *truth value*. 'Ted Sider is a philosopher and Barack Obama is a politician' is true; the permuted description 'Barack Obama is a philosopher and Ted Sider is a politician' is false; how then could they be equivalent? Similarly, where $\mathscr{C}$ is the nomic role played by charge and $\mathscr{M}$ is the nomic role played by mass, it's hard to see how 'charge plays $\mathscr{C}$ and mass plays $\mathscr{M}$' could be equivalent to 'charge plays $\mathscr{M}$ and mass plays $\mathscr{C}$' when the former is true and the latter is false.[44]

[42] Perhaps Mundy (1992) can be read in this way.

[43] McKenzie (2017) begins a survey article by saying 'While a number of distinct positions go under the banner of "ontic structural realism" (OSR), common to them is the insistence that the structural features of reality should be accorded ontologically fundamental status'.

[44] Thanks to Jeremy Goodman and others here. For similar reasons, one cannot articulate the structuralist thought in these permutational examples as the claim that the sentences are equivalent in the senses of the modal or higher-order accounts of equivalence from section 5.1.

This obstacle can be overcome, albeit at some cost in complexity. There is a defensible quotienting account of equivalence between merely permutationally distinct possibilities, but it requires introducing a new sense of equivalence.

Sentences that are equivalent in the sense we have been discussing so far in this chapter, such as '*a* is 1,000 grams' and '*a* is 1 kilogram', are equally good descriptions of the way things actually are. Call this sense *actual-equivalence*. 'Sider is a philosopher and Obama is a politician' and 'Obama is a philosopher and Sider is a politician' are *not* equally good descriptions of the way things actually are, since the first is true and the second is false. Thus those sentences are not actual-equivalent; actual-equivalent sentences must share truth value. Nevertheless some insist that there is no genuine difference between the actual world (in which the first sentence is true) and a qualitatively identical possible world in which Obama and I have exchanged roles (in which the second sentence is true). A quotiener can put this by saying that descriptions of these possible scenarios are equivalent in a new sense, which we may call *complete counterfactual* equivalence.

Similarly, even though 'charge plays $\mathscr{C}$ and mass plays $\mathscr{M}$' and 'charge plays $\mathscr{M}$ and mass plays $\mathscr{C}$' differ in truth value, a quotiener might claim that a complete description of the actual world is complete-counterfactual-equivalent to a complete description of a world in which charge and mass have exchanged roles.[45] For one further example, a quotiener who thinks that a global doubling of mass is a distinction without a difference could put this idea by saying that a complete description of actuality in which mass is described numerically, in terms of some chosen unit, is complete-counterfactual-equivalent to an exactly similar description but in which all numerical mass values are doubled.

Both parts of the name 'complete counterfactual' are significant. Complete: the new sort of equivalence does not hold between partial descriptions of possible worlds, but only between sentences taken as descriptions of entire possible worlds. There clearly *are* possible worlds genuinely distinct from the actual world in which I am a politician and Obama is a philosopher; it is only when one is considering a complete description of actuality, *A*, and a complete description *A'* of the world in which Obama and I have exchanged roles, where one is tempted to say that the descriptions pick out the same possibility. Counterfactual: the descriptions *A* and *A'* in fact differ in truth value. (*A* is true since it is stipulated to be a complete description of actuality; *A'* is false since it includes the false sentence 'Sider is a politician'.) Thus as before they are inequivalent if taken as descriptions of actuality. It is only when we consider one of them as a description of a counterfactual scenario that there is any temptation to claim equivalence.

The distinction between actual and complete counterfactual equivalence is perhaps not unintuitive. But is it ultimately intelligible? What is the difference?

---

[45] Assuming that both descriptions are in our language. If the latter were instead in a language in which 'mass' means charge and 'charge' means mass (and is otherwise just like ours) then that description would, in its language, be actual-equivalent to the former description in our language.

Here we run up against a problem. It would be nice to distinguish between the two by defining each of them and inspecting how the definitions differ; but the quotienter can't be expected to do that since the whole point of quotienting is to deny that claims of equivalence need definition.

A quotienter might simply rest with the intuitive explanation of the distinction above. But a certain indirect approach is available for saying something more. A quotienter could first say how the distinction could be defined by a friend of fundamentality, and then kick the ladder away. To some this indirect approach will be unsatisfying, but perhaps these will be the very philosophers (like me) who find quotienting in general unsatisfying.

I will develop the indirect approach in some detail, in the case of the putative complete counterfactual equivalence between statements that differ over a permutation of individuals. Let us consider how that case would look from a fundamentalist perspective, and in particular from the perspective of "quantifier generalism" (section 3.14), according to which fundamental facts are expressed using name-free sentences of predicate logic.

Given quantifier generalism, how do sentences involving proper names connect to fundamental reality? One possible answer would be that such sentences express propositions that are grounded by purely qualitative propositions. But there are serious objections to this approach. Consider, for example, Max Black's (1952) world consisting solely of two duplicate spheres, Castor and Pollux. As Robert Adams (1979) says, surely Castor could have been destroyed while Pollux lived on; and similarly, Pollux could have been destroyed while Castor lived on. If the proposition that Pollux is destroyed is in the first world grounded in some true qualitative propositions, those propositions would also be true in the second; but that would mean that in that second, Pollux is destroyed.[46]

A more attractive approach for a generalist is based on Dasgupta's (2020) approach to a similar problem for relationalism about chirality, quantity, and space.[47] On this approach statements containing proper names are not factual.

---

[46] On issues in this vicinity see Dasgupta (2014a); Russell (2017; 2018).

[47] When transposed to the case of individuals, this approach is more like his original approach in Dasgupta (2009) than the one in Dasgupta (2014a). (Sider (2008b) is in a similar spirit. Russell (2015) also defends a related view; the view to be presented can be thought of as the analogue, in the formal mode, of what is in the material mode in Russell.) Incidentally, Dasgupta (2009, section 4.4) presents an attractive rationalization of our practice of using proper names, along these lines:

> Communication with purely qualitative sentences is inefficient. For suppose one wants to report a discovery that a previously familiar object has some further feature. In purely qualitative terms, the initial knowledge is that 'some object has features $F_1, \ldots, F_n$', and the discovery is that 'some object has features $F_1, \ldots, F_n$ and also F_{n+1}'. Thus all the prior knowledge would need to be reiterated in the report. But with names this transition can be streamlined. The commitment to the initial knowledge can be made by uttering 'N has F_1", $\ldots$, "N has F_n'. And the discovery can then be reported with 'N has F_{n+1}', since this together with the earlier sentences implies a commitment to there being some object with features $F_1, \ldots, F_n$ and also F_{n+1}.

They are incapable of truth or falsity, they do not express propositions, and as a result are not grounded in propositions about fundamental reality. They can, however, be aptly uttered; and although there are no sufficient conditions, stated in fundamental terms, for aptness, there is a necessary condition for the aptness of any collection of statements: that its ramsey sentence be true. That is, where $S_1(\alpha_1, \alpha_2, \dots), S_2(\alpha_1, \alpha_2, \dots), \dots$ are sentences containing names $\alpha_1, \alpha_2, \dots$ (perhaps not all of them occurring in each sentence), the entire collection is apt only if there exist $x_1, x_2, \dots$ such that $S_1(x_1, x_2, \dots), S_2(x_1, x_2, \dots), \dots$.

The function of the necessary condition is *coordination*. Uses of a name must be coordinated with other uses of the name—both past and present, both solo and in conjunction with other names, both by oneself and by other speakers. The coordination consists in the truth of the ramsey sentence for the set of all the uses.

Realistically, though, it just isn't true that *all* uses of sentences involving names (across times, by different speakers) are coordinated in this sense. We need a more localized, practical notion of aptness, applicable even in the case of an utterance of a single sentence S by some speaker.[48] For this more localized notion we cannot look merely to the ramsey sentence of S. An utterance by me of 'Barack Obama was born in Kenya' is not apt, but its ramsey sentence 'Someone was born somewhere' is true. We must choose some appropriate set of "background" sentences (uttered perhaps by the speaker, perhaps by others, perhaps even by no one)—expansive enough to adequately constrain the names in S, but not so expansive as to make aptness unachievable—and say that the utterance is apt only if S plus the background has a true ramsey sentence.

Consider, now, statements of equivalence involving proper names. (Dasgupta himself wouldn't endorse the following. Remember that we are exploring a ladder for the quotienter to kick away. Dasgupta is no quotienter.) There is a natural way to extend the account of actual equivalence from section 5.3 to the present context. Let S and S' be two sentences containing proper names, and let B be an appropriate background for those sentences. We can then say that S and S' are actual-equivalent if and only if the ramsey sentence of $S + B$ (that is, $\{S\} \cup B$) is actual-equivalent, in the sense of section 5.3, to the ramsey sentence of $S' + B$—that is, if and only if those two ramsey sentences "say the same thing about fundamental reality". (Note that this counts the statement that S and S' are actual-equivalent as being capable of truth, not just aptness.) Actual-equivalent sentences make the

However, this presentation (which is mine, not Dasgupta's) assumes an overly qualitative epistemology, in that we begin with qualitative knowledge and move to a qualitative conclusion, the only question being how to report this concisely. But even setting aside controversies about the semantics of proper names in natural language, presumably our initial and final knowledge, and also our evidence, is all nonqualitative, because largely *de se*. The rationalization should be more thoroughgoing: since communication and thought in a purely qualitative language would be ineffecient, we think and speak nonqualitatively from start to finish; it's just that the metaphysical underpinnings of this talk and thought are purely qualitative.

[48] Thanks to John Hawthorne here.

same fundamental contribution, given the background, to the apt representation of the actual qualitative facts.

To illustrate, where c is a proper name of a certain massive object, consider:

(1) c is 1 kg
(2) c is 1,000 g.

Relative to any background, B, the ramsey sentences for (1) + B and (2) + B are, respectively,

$$\exists x \dots (x \text{ is } 1 \text{ kg} \wedge B(x, \dots))$$
$$\exists x \dots (x \text{ is } 1{,}000 \text{ g} \wedge B(x, \dots)).$$

Assuming these say the same thing about fundamental reality, (1) and (2) count as actual equivalent.

But now consider:

(SO) Sider is a philosopher and Obama is a politician
(OS) Obama is a philosopher and Sider is a politician

relative to a background B that includes these sentences:

(S) Sider is not a politician
(O) Obama is not a philosopher.

(SO) and (OS) would *not* be actual equivalent, for the ramsey sentences of (SO) + B and (OS) + B do not say the same thing about fundamental reality:

$$\exists x \exists y \dots (x \text{ is a philosopher} \wedge y \text{ is a politician} \wedge x \text{ is not a politician} \wedge y \text{ is not a philosopher} \wedge \dots)$$
$$\exists x \exists y \dots (x \text{ is a philosopher} \wedge y \text{ is a politician} \wedge y \text{ is not a politician} \wedge x \text{ is not a philosopher} \wedge \dots).$$

(The latter is contradictory.) In order for (SO) and (OS) to be actual equivalent, what they require of fundamental reality *together with the background* would need to be the same.

Next let us turn to complete counterfactual equivalence. Recall the intuitive idea: descriptions A and A' are complete counterfactual equivalent when there is no genuine difference between a complete possibility described by A and a complete possibility described by A'. And in the present context, it is natural to understand this as requiring that the complete *fundamental* story of the two possibilities be the same.[49] And these complete fundamental stories are simply the ramsey sentences of A and A'. Thus if A is a complete description of actuality and A' is the result of permuting 'Ted Sider' and 'Barack Obama' throughout A, A and A' have the same ramsey sentence and hence are complete-counterfactual-equivalent.

Thus the crucial difference between actual and complete counterfactual equivalence is that the latter requires no coordination with the background. In actual

---

[49] Dasgupta allows a "loose" sense of possibility on which there are distinct such possibilities, all corresponding to the same "strict" possibility—that is, the same "arrangement of fundamental matters". In these terms, complete-counterfactual-equivalent sentences correspond to the same strict possibility.

equivalence one is taking the sentences as if they were to be asserted about actuality, and so coordination with the background is required. But in complete counter-factual equivalence, one is not taking the sentences as if they were assertions about actuality, but rather as complete specifications of alternate scenarios. No coordination with the background is then required; the only question is whether the alternate scenarios are fundamentally the same. In the case of descriptions that differ by a mere permutation, these two kinds of equivalence come apart.

That, then, is the ladder, which the quotienter must ultimately kick away. From *my* perspective—a hostile perspective—I can now see what quotienters are trying to get at, despite themselves. In claiming that descriptions A and A' of permutationally distinct worlds are complete-counterfactual-equivalent, though not actual-equivalent, they are in effect assuming that fundamental reality is as the quantifier generalist says, and saying that although we cannot aptly assert both A and A' (since they do not both coordinate with the background), they nevertheless correspond to the same fundamental possibility: they share the same ramsey sentence. But quotienters will need to insist that their account is intelligible without this further perspective.

5.8 Against quotienting

What can be said in defence of the fundamentality approach to equivalence, and against quotienting? I don't have anything decisive to say. My main goal is to get the issues here out in the open: the possibility of quotienting, and the metaphysician's usual presupposition of the requirement of artefact-free representation. Still, I do oppose quotienting and will say what I can against it.

5.8.1 *Quotienting is unsatisfying*

In actual fact, quotienting isn't normally pursued—except when there seems to be no other choice. Philosophers of science, for example, have invested a lot of effort in finding representation theorems for various sorts of measurement. Granted, this work is motivated by many different desires, but surely part of it is dissatisfaction with merely quotienting out the conventional content of a unit of measurement—a desire for a more satisfying account of *why* any scalar transformation of a mass function is just as good as any other. Similarly, philosophers of physics often prefer coordinate-free formulations of geometric theories, rather than coordinate formulations with quotiented-out conventional content.

This is especially evident when we consider ontology. The following attitude is *not* typical: 'the facts in a certain domain can be well represented by a theory that takes there to exist certain objects, but we could equivalently take there to exist quite different objects provided we make certain adjustments'. A typical reaction to this attitude would be that until an ontology is specified, it hasn't been made clear what the theory is saying about the world. (That's not to say that there aren't exceptions—David Wallace is an example.)

Why isn't quotienting normally pursued? The answer is right at the surface: it is intuitively unsatisfying to give no answer to the question of why relations of equivalence hold. When multiple equally good ways to represent are available, it is natural to ask why that is, to ask what it is about reality that enables it to be multiply represented.

Consider Leibniz himself. His relationalism would have been far less compelling if he had said merely that descriptions of objects in space are equivalent when and only when they differ only by some combination of global translations and rotations, rather than explaining that equivalence, as he did, with a claim about fundamental reality—that it contains material bodies standing in spatial relations, rather than points of substantival space.

Most will, I think, be sympathetic with the previous paragraph; they will be happy to demand some explanation of the equivalence. But fewer will be happy with the full demand I would make: I would ask of Leibniz not only that he specify the ontology of his theory that generates the equivalence (material bodies), and the *kinds* of properties and relations the theory concerns (spatial relations), but also that he specify *which* spatial relations are the fundamental ones—are the fundamental relations the Tarskian (1999) relations $\equiv$ and B of equidistance and collinearity, or some others? It's interesting that most philosophers' intuitions are thus halfway between our two extreme approaches.

I demand explanations of relations of equivalence; quotienters reject this demand. Can we make any dialectical progress? Perhaps. Suppose quotienters acknowledge in *some* cases that it's better to explain relations of equivalence in terms of a deeper theory. Suppose, for instance, they concede that moving from a family of unit-based theories of mass (together with an appropriate equivalence relation) to a single theory based on $\succeq$ and C is some sort of theoretical improvement, or that coordinate-free geometric theories are preferable to coordinate-based ones, and that this is for metaphysical reasons. Then a tension in their position has arisen, since they can no longer maintain that there is nothing at all wrong with quotienting. They may reply that quotienting is a necessary evil: to be avoided whenever possible, but sometimes unavoidable. But in so saying they concede that what *I* say should be the goal—a theory in need of no quotienting—would indeed be superior, if it could be achieved.

Many writers on the philosophy of physics occupy a potentially unstable middle ground regarding quotienting. On one hand they're uncomfortable with full-on quotienting, if they take seriously the demand to say something specific about the ontology of space, for instance, or about the metaphysics of the wave function. But on the other hand, they don't take all the questions of the metaphysician seriously, for instance questions in the metaphysics of quantity about which predicates exactly generate quantitative facts.

My argument with the quotienter is an instance of a larger pattern. Foes of metaphysics accuse metaphysicians of asking questions whose weight our language cannot bear: language has clear sense when used within certain contexts (such as within science) but not otherwise. Friends reply that it is difficult to allow the familiar questions while disqualifying metaphysical ones. I myself regard the demand for an explanation of equivalence in terms of third languages as akin to the explanatory demands that motivate physics: in each case we seek to understand how phenomena arise from reality's ultimate constituents. But foes of metaphysics will regard the demand for metaphysical understanding as being a perversion of the legitimate demand for scientific understanding. This overarching opposition is ongoing.

5.8.2 *Progress can be unexpected*

I turn now to defensive manoeuvres on behalf of the fundamentality-based approach.

As we saw, the approach leads to problematic questions, for example whether the fundamental quantificational notion is ∀ or ∃. One concern about such questions is that they seem unanswerable. Worse, it seems as though there couldn't possibly be any considerations that would even slightly favour one answer as opposed to another.

But sometimes we do, in the end, make progress on questions that initially seemed hopeless. There is an excellent recent example of this. Kit Fine's 'Towards a Theory of Part' gives compelling—which is not to say irresistible—reasons to favour basing mereology on fusion rather than parthood or overlap. Such questions are, we would have antecedently thought, paradigmatically unanswerable, and indeed questions on which one could *never* have the *slightest* reason in favour of one answer or another. But then along comes Fine giving reasons—of a completely different sort than I, anyway, had expected to be relevant—for a particular answer to the question. It can happen. We can't know in advance that it won't.

It's worth remembering that 'we could never possibly have reason to favour one answer or another' is precisely what nonphilosophers tend to think about all of philosophy. It's practically the job description of a philosopher to somehow find considerations relevant to what look initially like questions that are good only for baseless speculation.

5.8.3 *Hard choices are hard to avoid*

The worrisome thing is being saddled with the choice of whether to say that, for example, it is ∃ or ∀ that is fundamental. But for most of us, *some* such choices are inevitable. Recall Goodman's grue and bleen. We can play the same trick with fundamental physical quantities, and rewrite physical theories using the cooked-up predicates. Now, if you're completely happy with quotienting, you will say that there's nothing wrong with this, so long as we specify the relevant relations of equivalence. But suppose you're at least somewhat sympathetic to the fundamentality approach, and admit that the "grueified" theories fail to properly express the physical facts. The objects attracted each other because they had opposite charges, you insist, and not because they had opposite schmarges and were first observed before 3000 AD! In that case you admit *some* questions of the ∃ versus ∀ variety; and if you don't like the ∃ vs ∀ question itself, the problem becomes where to draw the line.

5.8.4 *There can be more than one*

Another defensive move: the question 'is it ∃ or ∀ that is fundamental?' leaves out a third possibility, namely that they are *both* fundamental.

I grant a general presumption in favour of fewer fundamental concepts—that's one sort of parsimony. But parsimony isn't the only relevant consideration: there is also avoidance of arbitrariness. Parsimony should sometimes be sacrificed to avoid arbitrariness.[50]

5.8.5 *Why think we can know everything?*

Pointing out that 'both' is an available answer to 'is it ∃ or ∀ that is fundamental?' mitigates one concern, which is that there should be no metaphysical asymmetry between ∃ and ∀. But another concern is simply that we have no way of knowing what the answer to the question is. That concern remains, since we seem to have no way of knowing whether it's ∃ or ∀ or both that is fundamental. Yes, unexpected progress is always possible, but sometimes, I'll admit, this seems very unlikely.

But even in such cases, there is the simple realist reply: why think we can know everything? Defenders of the fundamentality approach should not be shy about saying that they do not always know which concepts are fundamental. L. A. Paul (2012b, p. 21) put it well: 'It is the fate of philosophy to have many too many options'.

Though that is the central reply, more can be said to make that reply feel less abrupt, and more comfortable to embrace.

First, I really do think that philosophers (nonmetaphysicians, mostly) throw around the 'your proposed metaphysics leads to unknowable facts, so it should be rejected' argument far too easily. Attempting to seriously defend such an argument lands one in a very old dialectic: one needs a principle that bans the target metaphysical questions without also eliminating, for example, legitimate scientific questions.

In my view, unknowability does not *on its own* constitute something's having gone wrong, but it might be a sign that the theory is employing concepts that aren't in good standing. We need to look at the concepts involved in the unknowable question, to see whether there is good reason, on general systematic grounds, to think that those concepts are in good standing. In particular, do those concepts play a central role in *other* questions that are part of legitimate inquiry? If so, that is a good reason to think that they are in good standing; and then, the fact that they can also be used to raise a question we can't see how to answer (or even

---

[50] See Sider (2011, section 10.2).

begin answering) is no reason at all to think that the question is somehow ill-posed. Imagine that the empirical evidence favours a theory that posits particles. We could then raise various questions, using the concepts of this theory, that are in a practical sense unanswerable by creatures like us, such as whether the number of particles in the entire universe is even or odd. Such questions are clearly legitimate because they're framed using concepts we have reason to think are in good standing—concepts that latch onto real features of the world. The fact that those concepts can be combined to raise a question that is practically unanswerable has no tendency at all to show that the question is illegitimate. Nor is it clear why unanswerability in a more in-principle sense would have any greater significance.

Second, someone who rejects the concept of fundamentality because of a desire to avoid unanswerable questions may be living in a glass house. For rejecting the notion of fundamentality is itself a metaphysical stance, albeit a negative one, and it's not at all clear what justifies it.

6

The Fundamentalist Vision

The postmodal thought is, at bottom, an explanatory one. If a truth is necessary, there must be some reason for why this is so.

This demand for explanation collides with certain forms of structuralism. I myself take the collision to undermine those structuralisms; committed structuralists may draw the opposite moral. Structuralism has often been formulated modally: nodes in a structure cannot vary independently from the way the nodes are structured. But postmodalism demands a deeper explanation of such necessary connections. And when the structuralism in question concerns indispensable elements of foundational science and mathematics, such as properties (Chapter 2) and individuals (Chapter 3), and, further, concerns the very existence and identities of those elements (as opposed to merely concerning certain features of them), it can be unclear what that deeper explanation—the deeper statement of structuralism—might be. The deeper statement must somehow privilege the structure and downgrade the nodes. But if it is the very existence and identity of the nodes that is at issue, the obvious way to downgrade the nodes is to eliminate them, deny that they exist; and it is difficult to formulate a coherent view of this sort, as we saw in our discussion of structural realism. And we cannot simply shift to some other structure based on different nodes, if the nodes in question are indispensable elements of scientific theorizing.

(Comparativism about quantity (Chapter 4) is not a structuralism of this sort, for there we *can* shift to a structure based on different nodes: we can shift from a structure whose nodes are absolute values of quantities, standing in higher-order comparative relations, to a structure whose nodes are concrete individuals, standing in first-order comparative relations. The distinctively postmodal problems faced by comparativism—many of them shared by various forms of absolutism— are different from those faced by nomic essentialism and structuralism about individuals.)

Though not all postmodalists will agree, the postmodalist demand for explanations of modal truths can be seen as a special case of something more general, which we might call the fundamentalist vision. According to this vision, there is such as thing as how the world fundamentally is—there is, that is, such a thing as what is ultimately going on—and much of what we are up to in metaphysics is seeking explanations of phenomena in terms of what is fundamental.

The fundamentalist vision led us to go beyond the postmodalist demand for explanations of modal truths, and to demand deeper explanations even for some postmodal claims. In contexts where we are trying to limn the ultimate structure of the world, I argued in sections 2.3 and 3.11, our final proposals cannot be essentialist or higher-level grounding claims, for such claims are in a sense metaphysically unspecific; we must instead say what is fundamental. Thus in these contexts the proper postmodal tool is fundamentality.

This fundamentalist vision leads naturally, though perhaps not inevitably, to a view of laws of physics: that "fundamentalist" laws of physics—laws of physics stated in a language all of whose expressions express fundamental concepts—can be found which have the scientific virtues we prize in our ordinary, less metaphysically loaded, pursuit of laws. On this view we can hope to find fundamentalist laws that have not only superempirical virtues such as simplicity and nonarbitrariness, but also more empirical virtues such as determinism (if that is warranted by the evidence). The picture here is that ordinary scientific methods of theory choice retain their applicability at the fundamental level.

None of this rich fundamentalist vision is inevitable. The whole postmodalist approach may be rejected. Some postmodalists will reject the emphasis on fundamentality. And friends of fundamentality may reject the presumption of fundamentalist laws. Still, I think the attractions of this vision are many, obvious, and powerful. I hope many philosophers of physics will recognize it as an articulation and generalization of something they have in effect presupposed all along, when they told themselves and others that philosophy of physics is the investigation of what physical theories are telling us about the fundamental nature of the physical world. This is not the only conception of philosophy of physics, and not even the only "realist" one, but perhaps it is the one most in line with the intuitive basis of realist thought about physics.

To my mind, the most serious challenges facing the fundamentalist vision are those having to do with arbitrariness. The first of these challenges confronts the fundamentalist vision of laws. As we saw in Chapter 4, a fundamentalist approach to quantities leads to embracing some nonnumeric conception of the fundamental quantitative concepts. But it is then difficult to avoid a certain artificiality in the laws of physics (section 4.7). Worse, fundamentalist laws can turn out indeterministic, even in contexts where ordinary scientific evidence favours determinism. Only by embracing laws based on certain arbitrary decisions can determinism be reinstated (sections 4.8–4.11). The second challenge is to the fundamentalist vision proper. As we saw in Chapter 5, the defender of that vision seems forced to regard what seem to be equivalent notational variants as instead being genuine alternatives. Again there is a threat of arbitrariness, though of a different sort: it seems arbitrary that one rather than another of these alternatives should be true.

In light of these challenges, the approach of "quotienting" took on importance, in diametric opposition to the fundamentalist vision. Two sorts of quotienting arose. One, quotienting proper, opposes the goal of saying, in one fixed way, what the fundamental truth about the world is. According to the quotienter, there can be many equally good ways of describing the world, even if there is no way to explain why these different ways are equally good—even if, that is, there is no way to give a privileged description of their common subject matter, no way of saying what is "really going on" (section 5.5). A second sort, nomic quotienting, also opposes the goal of giving privileged theories, but does so only for the laws of nature. While the nomic quotienter may grant (against the quotienter proper) that there is a privileged way to state the fundamental facts, she denies that there needs to be any single, privileged way to say, about those fundamental facts, what is nomically fixed. There can be many equally good ways to state laws of nature, and there need not be any explanation of this fact in more nomically basic terms (section 4.9.3).

Both forms of quotienting are, at bottom, the rejection of explanatory demands. When there are many equally good ways to formulate the facts, or the laws of nature, quotienters deny the need to explain this plurality in terms of a single, privileged claim. This is the polar opposite of the fundamentalist vision, which is based on embracing such explanatory demands. How far explanatory demands can legitimately be pressed is a primordial choice point, with far-reaching ramifications, in the metaphysics of science.

And it is deeply intertwined with the question of the proper conceptual tools for metaphysics. The tools matter.

References

Adams, Robert Merrihew (1979). Primitive thisness and primitive identity. *Journal of Philosophy*, 76, 5–26.

Albert, David Z. (1996). Elementary quantum metaphysics. In *Bohmian Mechanics and Quantum Theory: An Appraisal* (edited by J. T. Cushing, A. Fine, and S. Goldstein), pp. 277–84. Kluwer Academic Publishers, Dordrecht.

Albert, David Z. (2000). *Time and Chance*. Harvard University Press, Cambridge, MA.

Albert, David Z. (2015). Quantum mechanics and everyday life. In *After Physics*, pp. 124–43. Harvard University Press, Cambridge, MA.

Allori, Valia (2022a). Fundamental objects without fundamental properties: a thin-object-orientated metaphysics grounded on structure. In *Probing the Meaning and Structure of Quantum Mechanics* (edited by D. Aerts, J. Arenhart, C. De Ronde, and G. Sergioli). World Scientific. Forthcoming.

Allori, Valia (2022b). Towards a structuralist elimination of quantum properties. In *Quantum Mechanics and Fundamentality: Naturalizing Quantum Theory between Scientific Realism and Ontological Indeterminacy* (edited by Valia Allori). Springer. Forthcoming.

Anscombe, G. E. M. (1964). Substance. *Aristotelian Society, Supplementary Volume*, 38, 69–78.

Armstrong, D. M. (1980). Against "ostrich" nominalism. *Pacific Philosophical Quarterly*, 61, 440–9.

Armstrong, D. M. (1983). *What is a Law of Nature?* Cambridge University Press, Cambridge.

Armstrong, D. M. (1989). *A Combinatorial Theory of Possibility*. Cambridge University Press, New York.

Armstrong, D. M. (1997). *A World of States of Affairs*. Cambridge University Press, Cambridge.

Armstrong, D. M. (2004). *Truth and Truthmakers*. Cambridge University Press, Cambridge.

Arntzenius, Frank (2000). Are there really instantaneous velocities? *The Monist*, 83, 187–208.

Arntzenius, Frank (2012). *Space, Time, and Stuff*. Oxford University Press, Oxford.

Arntzenius, Frank and Dorr, Cian (2011). Calculus as Geometry. Chapter 8 of Frank Arntzenius, *Space, Time, and Stuff*. Oxford: Oxford University Press.

Bacon, John (1985). The completeness of a predicate-functor logic. *Journal of Symbolic Logic*, 50, 903–26.

Bader, Ralf M. (2019). Conditional grounding. Manuscript.

Baker, Alan (2016). Simplicity. Stanford Encyclopedia of Philosophy. Available at <https://plato.stanford.edu/archives/win2016/entries/simplicity>.

Baker, David John (2020). Some consequences of physics for the comparative metaphysics of quantity. In Bennett and Zimmerman (2020), pp. 75–112.

Barker, Stephen (2013). The emperor's new metaphysics of powers. *Mind*, 122, 605–53.

Barnes, Elizabeth (2014). Going beyond the fundamental: Feminism in contemporary metaphysics. *Proceedings of the Aristotelian Society*, 114, 335–51.

Barnes, Elizabeth (ed.) (2016). *Current Controversies in Metaphysics*. Routledge.

Barrett, Thomas William and Halvorson, Hans (2016a). Glymour and Quine on theoretical equivalence. *Journal of Philosophical Logic*, 45, 467–83.

Barrett, Thomas William and Halvorson, Hans (2016b). Morita equivalence. *Review of Symbolic Logic*, 9, 556–82.

Barrett, Thomas William and Halvorson, Hans (2017). From geometry to conceptual relativity. *Erkenntnis*, 82, 1043–63.

Belot, Gordon (2012). Quantum states for primitive ontologists: A case study. *European Journal for Philosophy of Science*, 2, 67–83.

Belot, Gordon (2013). Symmetry and equivalence. In *The Oxford Handbook of Philosophy of Physics* (edited by Robert Batterman), pp. 318–39. Oxford University Press, New York.

Benacerraf, Paul (1965). What numbers could not be. *Philosophical Review*, 74, 47–73.

Bennett, Karen (2017). *Making Things Up*. Oxford University Press, Oxford.

Bennett, Karen and Zimmerman, Dean W. (2013). *Oxford Studies in Metaphysics*, volume 8. Oxford University Press, Oxford.

Bennett, Karen and Zimmerman, Dean W. (2020). *Oxford Studies in Metaphysics*, volume 12. Oxford University Press, Oxford.

Bhogal, Harjit and Perry, Zee R. (2017). What the Humean should say about entanglement. *Noûs*, 51, 74–94.

Bigelow, John and Pargetter, Robert (1988). Quantities. *Philosophical Studies*, 54, 287–304.

Bird, Alexander (2007a). *Nature's Metaphysics: Laws and Properties*. Oxford University Press, Oxford.

Bird, Alexander (2007b). The Regress of Pure Powers? *Philosophical Quarterly*, 57, 513–34.

Bird, Alexander (2016). Overpowering: How the powers ontology has overreached itself. *Mind*, 125, 341–83.

Black, Max (1952). The identity of indiscernibles. *Mind*, 61, 153–64.

Blackburn, Simon (1990). Filling in space. *Analysis*, 50, 62–5.

Bliss, Ricki and Trogdon, Kelly (2016). Metaphysical grounding. Stanford Encyclopedia of Philosophy. Available at <https://plato.stanford.edu/archives/win2016/entries/grounding/>.

Boolos, George (1984). To be is to be the value of a variable (or to be some values of some variables). *Journal of Philosophy*, 81, 430–49.

Boolos, George (1985). Nominalist platonism. *Philosophical Review*, 94, 327–44.

Bricker, Phillip (1992). Realism without parochialism. Read to the Pacific Division of the APA. Available at <http://blogs.umass.edu/bricker/files/2013/08/realism_without_parochialism.pdf>.

Brighouse, Carolyn (1994). Spacetime and Holes. *PSA: Proceedings of the Biennial Meeting of the Philosophy of Science Association*, 1994, 117–25.

Burgess, Alexis (2012). A puzzle about identity. *Thought*, 1, 90–9.

Butterfield, Jeremy (1989). The hole truth. *British Journal for the Philosophy of Science*, 40, 1–28.

Campbell, Keith (1990). *Abstract Particulars*. Blackwell, Oxford.

Cartwright, Nancy (1999). *The Dappled World: A Study of the Boundaries of Science*. Cambridge University Press, Cambridge.

Chakravartty, Anjan (2012). Ontological priority: The conceptual basis of non-eliminative, ontic structural realism. In *Structural Realism: Structure, Object, and Causality* (edited by Elaine Landry and Dean Rickles), pp. 187–206. Springer, Dordrecht.

Chalmers, David J., Manley, David, and Wasserman, Ryan (2009). *Metametaphysics*. Oxford University Press, Oxford.

Chen, Eddy Keming (2017). Our fundamental physical space: An essay on the metaphysics of the wave function. *Journal of Philosophy*, 114, 333–365.

Cohen, Shlomit Wygoda (2020). Not all partial grounds partly ground: Some useful distinctions in the theory of grounding. *Philosophy and Phenomenological Research*, 100, 75–92.

Correia, Fabrice (2013). Logical grounds. *Review of Symbolic Logic*, 7, 1–29.

Cross, Troy (2004). The nature of fundamental properties. PhD thesis, Rutgers University, NJ.

Cross, Troy (2012a). Goodbye, Humean supervenience. In *Oxford Studies in Metaphysics* (edited by Karen Bennett and Dean W. Zimmerman), volume 7, pp. 129–53. Oxford University Press, Oxford.

Cross, Troy (2012b). Review of Marmodoro, *Metaphysics of Powers: Their Grounding and Their Manifestations*. *Notre Dame Philosophical Reviews*.

Dasgupta, Shamik (2009). Individuals: An essay in revisionary metaphysics. *Philosophical Studies*, 145, 35–67.

Dasgupta, Shamik (2011). The bare necessities. *Philosophical Perspectives*, 25, 115–60.

Dasgupta, Shamik (2013). Absolutism vs comparativism about quantity. In Bennett and Zimmerman (2013), pp. 105–48.

Dasgupta, Shamik (2014a). On the plurality of grounds. *Philosophers' Imprint*, 14, 1–28.

Dasgupta, Shamik (2014b). The possibility of physicalism. *Journal of Philosophy*, 111, 557–92.

Dasgupta, Shamik (2015a). Inexpressible ignorance. *Philosophical Review*, 124, 441–80.

Dasgupta, Shamik (2015b). Substantivalism vs relationalism about space in classical physics. *Philosophy Compass*, 10, 601–24.

Dasgupta, Shamik (2016a). Can we do without fundamental individuals? Yes. In Barnes (2016), pp. 7–23.

Dasgupta, Shamik (2016b). Symmetry as an epistemic notion (twice over). *British Journal for the Philosophy of Science*, 67, 837–78.

Dasgupta, Shamik (2018). Realism and the absence of value. *Philosophical Review*, 127, 279–322.

Dasgupta, Shamik (2020). How to be a relationalist. In Bennett and Zimmerman (2020), pp. 113–63.

Dasgupta, Shamik (2021). Symmetry and superfluous structure: A metaphysical overview. In *The Routledge Companion to Philosophy of Physics* (edited by Eleanor Knox and Alastair Wilson), pp. 551–61. Routledge, New York.

Dasgupta, Shamik and Turner, Jason (2016). Postscript. In Barnes (2016), pp. 35–42. Postscript to Dasgupta (2016a) and Turner (2016a).

Dewar, Neil (2015). Symmetries and the philosophy of language. *Studies in History and Philosophy of Science Part B: Studies in History and Philosophy of Modern Physics*, 52, 317–27.

Dewar, Neil (2019a). Algebraic structuralism. *Philosophical Studies*, 176, 1831–54.

Dewar, Neil (2019b). Sophistication about symmetries. *British Journal for the Philosophy of Science*, 70, 485–521.

Dipert, Randall R. (1997). The mathematical structure of the world: The world as graph. *Journal of Philosophy*, 94, 329–58.

Donaldson, Thomas (2015). Reading the book of the world. *Philosophical Studies*, 172, 1051–77.

Dorr, Cian (2004). Non-symmetric relations. In *Oxford Studies in Metaphysics* (edited by Dean W. Zimmerman), volume 1, pp. 155–92. Oxford University Press, Oxford.

Dorr, Cian (2007). There are no abstract objects. In *Contemporary Debates in Metaphysics* (edited by Theodore Sider, John Hawthorne, and Dean W. Zimmerman), pp. 32–63. Blackwell, Oxford.

Dorr, Cian (2010a). Of numbers and electrons. *Proceedings of the Aristotelian Society*, 110, 133–81.

Dorr, Cian (2010b). Review of James Ladyman and Don Ross, *Every Thing Must Go: Metaphysics Naturalized. Notre Dame Philosophical Reviews*, 2010.

Dorr, Cian (2016). To be F is to be G. *Philosophical Perspectives*, 30, 39–134.

Dorr, Cian and Hawthorne, John (2013). Naturalness. In Bennett and Zimmerman (2013), pp. 3–77.

Dunn, J. Michael (1990). Relevant predication 3: Essential properties. In *Truth or Consequences* (edited by J. Dunn and A. Gupta), pp. 77–95. Kluwer Academic Publishers, Dordrecht.

Earman, John (1989). *World Enough and Space-Time: Absolute versus Relational Theories of Space and Time*. MIT Press, Cambridge, MA.

Earman, John and Norton, John (1987). What price spacetime substantivalism? the hole story. *British Journal for the Philosophy of Science*, 38, 515–25.

Eddon, M. (2011). Intrinsicality and hyperintensionality. *Philosophy and Phenomenological Research*, 82, 314–36.

Eddon, M. (2017). Fundamental properties of fundamental properties. In *Oxford Studies in Metaphysics* (edited by Karen Bennett and Dean W. Zimmerman), volume 10, pp. 78–104. Oxford University Press, Oxford.

Eklund, Matti (2006). Metaontology. *Philosophy Compass*, 1, 317–34.

Esfeld, Michael (2003). Do relations require underlying intrinsic properties? A physical argument for a metaphysics of relations. *Metaphysica: International Journal for Ontology and Metaphysics*, 4, 5–25.

Esfeld, Michael (2004). Quantum entanglement and a metaphysics of relations. *Studies in History and Philosophy of Science Part B*, 35, 601–17.

Esfeld, Michael (2017). A proposal for a minimalist ontology. *Synthese*, pp. 1–17.

Esfeld, Michael (2020). Super-Humeanism: The Canberra plan for physics. In *The Foundation of Reality: Fundamentality, Space and Time* (edited by David Glick, George Darby, and Anna Marmodoro), pp. 125–38. Oxford University Press, Oxford.

Esfeld, Michael and Deckert, Dirk-Andre (2017). *A Minimalist Ontology of the Natural World*. Routledge, New York.

Esfeld, Michael and Lam, Vincent (2008). Moderate structural realism about space-time. *Synthese*, 160, 27–46.

Esfeld, Michael and Lam, Vincent (2011). Ontic structural realism as a metaphysics of objects. In *Scientific Structuralism* (edited by Alisa Bokulich and Peter Bokulich), pp. 143–59. Springer Science+Business Media.

Esfeld, Michael, Lazarovici, Dustin, Hubert, Mario, and Dürr, Detlef (2014). The ontology of Bohmian mechanics. *British Journal for the Philosophy of Science*, 65, 773–96.

Evans, Matthew (2012). Lessons from *Euthyphro* 10a–11b. *Oxford Studies in Ancient Philosophy*, 42, 1–38.

Field, Hartry (1980). *Science Without Numbers*. Blackwell, Oxford.

Field, Hartry (1985). Can we dispense with spacetime? In *PSA 1984: Proceedings of the 1984 Biennial Meeting of the Philosophy of Science Association* (edited by P. Asquith and P. Kitcher), volume 2, pp. 33–90. Michigan State University Press, East Lansing.

Fine, Kit (1994a). Essence and modality. In Tomberlin (1994), pp. 1–16.

Fine, Kit (1994b). Senses of essence. In *Modality, Morality, and Belief* (edited by Walter Sinnott-Armstrong), pp. 53–73. Cambridge University Press, New York.

Fine, Kit (1995). Ontological dependence. *Proceedings of the Aristotelian Society*, 95, 269–90.

Fine, Kit (2000). Neutral relations. *Philosophical Review*, 109, 1–33.

Fine, Kit (2001). The question of realism. *Philosophers' Imprint*, 1, 1–30.

Fine, Kit (2005). Necessity and non-existence. In *Modality and Tense*, pp. 321–55. Oxford University Press, New York.

Fine, Kit (2007). Relatively unrestricted quantification. In *Absolute Generality* (edited by Agustín Rayo and Gabriel Uzquiano), pp. 20–44. Oxford University Press, Oxford.

Fine, Kit (2010a). Some puzzles of ground. *Notre Dame Journal of Formal Logic*, 51, 97–118.

Fine, Kit (2010b). Towards a theory of part. *Journal of Philosophy*, 107, 559–89.

Fine, Kit (2012). Guide to ground. In *Metaphysical Grounding: Understanding the Structure of Reality* (edited by Fabrice Correia and Benjamin Schnieder), pp. 37–80. Cambridge University Press, Cambridge.

Fine, Kit (2013). Fundamental truth and fundamental terms. *Philosophy and Phenomenological Research*, 87, 725–32.

Fine, Kit (2016). Identity criteria and ground. *Philosophical Studies*, 173, 1–19.

Forrest, Peter (1988). *Quantum Metaphysics*. Blackwell, Oxford.

Frege, Gottlob (1884). *The Foundations of Arithmetic*. Blackwell, Oxford, 2nd edition. Translated by J. L. Austin.

French, Steven (2010). The interdependence of structure, objects and dependence. *Synthese*, 175, 89–109.

French, Steven (2014). *The Structure of the World: Metaphysics and Representation*. Oxford University Press, Oxford.

French, Steven and Krause, Décio (2006). *Identity in Physics: A Historical, Philosophical, and Formal Analysis*. Oxford University Press, Oxford.

French, Steven and Ladyman, James (2003). Remodelling structural realism: Quantum physics and the metaphysics of structure. *Synthese*, 136, 31–56.

French, Steven and Redhead, Michael (1988). Quantum physics and the identity of indiscernibles. *British Journal for the Philosophy of Science*, 39, 233–46.

Fritz, Peter (2021). Ground and grain. *Philosophy and Phenomenological Research*. Forthcoming.

Gilmore, Cody (2013). Location and mereology. Stanford Encyclopedia of Philosophy. Available at <https://plato.stanford.edu/entries/location-mereology>.

Glazier, Martin (2016). Laws and the completeness of the fundamental. In *Reality Making* (edited by Mark Jago), pp. 11–37. Oxford University Press, Oxford.

Glymour, Clark (1970). Theoretical realism and theoretical equivalence. *PSA: Proceedings of the Biennial Meeting of the Philosophy of Science Association*, 1970, 275–88.

Glymour, Clark (1977). The epistemology of geometry. *Noûs*, 11, 227–51.

Gómez Sánchez, Verónica (2021). Naturalness by law. *Noûs*. Forthcoming.

Goodman, Jeremy (2017). Reality is not structured. *Analysis*, 77, 43–53.

Goodman, Nelson (1951). *The Structure of Appearance*. Harvard University Press, Cambridge, MA.

Goodman, Nelson (1955a). *Fact, Fiction, and Forecast*. Harvard University Press, Cambridge, MA.

Goodman, Nelson (1955b). The new riddle of induction. In Goodman (1955a), pp. 59–83.

Goodman, Nelson (1978). *Ways of Worldmaking*. Hackett, Indianapolis.

Greaves, Hilary (2011). In search of (spacetime) structuralism. *Philosophical Perspectives*, 25, 189–204.

Hale, Bob and Wright, Crispin (2001). *The Reason's Proper Study*. Oxford University Press, Oxford.

Hall, Ned (2015). Humean reductionism about laws of nature. Manuscript. Available at <https://philpapers.org/archive/HALHRA.pdf>.

Hawthorne, John (1995). The bundle theory of substance and the identity of indiscernibles. *Analysis*, 55, 191–6.

Hawthorne, John (2001). Causal structuralism. *Philosophical Perspectives*, 15, 361–78.

Hawthorne, John (2003). Identity. In *Oxford Handbook of Metaphysics* (edited by Michael J. Loux and Dean W. Zimmerman), pp. 99–130. Oxford University Press, Oxford.

Hawthorne, John and Sider, Theodore (2002). Locations. *Philosophical Topics*, 30, 53–76.

Healey, Richard (2007). *Gauging What's Real: The Conceptual Foundations of Contemporary Gauge Theories*. Oxford University Press, Oxford.

Hellman, Geoffrey (1989). *Mathematics without Numbers*. Oxford University Press, Oxford.

Hicks, Michael Townsen and Schaffer, Jonathan (2017). Derivative properties in fundamental laws. *British Journal for the Philosophy of Science*, 68, 411–50.

Hirsch, Eli (2002). Quantifier variance and realism. *Philosophical Issues*, 12, 51–73.

Hirsch, Eli (2011). *Quantifier Variance and Realism: Essays in Metaontology*. Oxford University Press, New York.

Hoefer, Carl (2016). Causal determinism. Stanford Encyclopedia of Philosophy. Available at <https://plato.stanford.edu/archives/spr2016/entries/determinism-causal/>.

Holton, Richard (1999). Dispositions all the way round. *Analysis*, 59, 9–14.

Horgan, Terence and Potr, Matjaž (2000). Blobjectivism and indirect correspondence. *Facta Philosophica*, 2, 249–70.

Horwich, Paul (1978). On the existence of time, space and space-time. *Noûs*, 12, 397–419.

Ismael, Jenann and van Fraassen, Bas C. (2003). Symmetry as a guide to superfluous theoretical structure. In *Symmetries in Physics: Philosophical Reflections* (edited by Katherine Brading and Elena Castellani), pp. 371–92. Cambridge University Press, Cambridge.

Jubien, Michael (1993). *Ontology, Modality, and the Fallacy of Reference*. Cambridge University Press, Cambridge.

Kang, Li (2019). Grounding, meet structuralism! Manuscript.

Kim, Jaegwon (1990). Supervenience as a philosophical concept. *Metaphilosophy*, 21, 1–27.

King, Jeffrey C. (2019). Structured propositions. Stanford Encyclopedia of Philosophy. Available at <http://plato.stanford.edu/entries/propositions-structured/>.

Kleinschmidt, Shieva (2014). *Mereology and Location*. Oxford University Press, Oxford.

Koslicki, Kathrin (2008). *The Structure of Objects*. Oxford University Press, Oxford.

Koslicki, Kathrin (2015). The coarse-grainedness of grounding. *Oxford Studies in Metaphysics*, 2015, 306–44.

Kovacs, David (2018). The deflationary theory of ontological dependence. *Philosophical Quarterly*, 68, 481–502.

Kovacs, David Mark (2017). Grounding and the argument from explanatoriness. *Philosophical Studies*, 174, 2927–52.

Krantz, David H., Luce, R. Duncan, Suppes, Patrick, and Tversky, Amos (1971a). *Foundations of Measurement*. Academic Press, New York. (3 volumes).

Krantz, David H., Luce, R. Duncan, Suppes, Patrick, and Tversky, Amos (1971b). *Foundations of Measurement, Vol. I: Additive and Polynomial Representations*. Academic Press, New York.

Kripke, Saul (1972). Naming and necessity. In *Semantics of Natural Language* (edited by Donald Davidson and Gilbert Harman), pp. 253–355, 763–9. D. Reidel, Dordrecht. Revised edition published in 1980 as *Naming and Necessity* (Harvard University Press, Cambridge, MA).

Ladyman, James (1998). What is structural realism? *Studies in History and Philosophy of Science*, 29, 409–24.

Ladyman, James (2014). Structural realism. Stanford Encyclopedia of Philosophy. Available at <https://plato.stanford.edu/entries/structural-realism/>.

Ladyman, James and Ross, Don (2007). *Every Thing Must Go: Metaphysics Naturalized*. Oxford University Press, Oxford. With David Spurrett and John Collier.

Langton, Rae (1998). *Kantian Humility: Our Ignorance of Things in Themselves*. Oxford University Press, Oxford.

Langton, Rae (2004). Elusive knowledge of things in themselves. *Australasian Journal of Philosophy*, 82, 129–36.

Lewis, David (1968). Counterpart theory and quantified modal logic. *Journal of Philosophy*, 65, 113–26.

Lewis, David (1971). Counterparts of persons and their bodies. *Journal of Philosophy*, 68, 203–11.

Lewis, David (1973). *Counterfactuals*. Blackwell, Oxford.

Lewis, David (1983). New work for a theory of universals. *Australasian Journal of Philosophy*, 61, 343–77.

Lewis, David (1984). Putnam's paradox. *Australasian Journal of Philosophy*, 62, 221–36.

Lewis, David (1986a). *On the Plurality of Worlds*. Blackwell, Oxford.

Lewis, David (1986b). *Philosophical Papers, Volume 2*. Oxford University Press, Oxford.

Lewis, David (1992). Critical notice of Armstrong's *A Combinatorial Theory of Possibility*. *Australasian Journal of Philosophy*, 70, 211–24.

Lewis, David (1994). Humean supervenience debugged. *Mind*, 103, 473–90.

Lewis, David (2009). Ramseyan humility. In *Conceptual Analysis and Philosophical Naturalism* (edited by David Braddon-Mitchell and Robert Nola), pp. 203–22. MIT Press, Cambridge, MA.

Linnebo, Øystein (2008). Structuralism and the notion of dependence. *Philosophical Quarterly*, 58, 59–79.

Linsky, Bernard and Zalta, Edward N. (1994). In defense of the simplest quantified modal logic. In Tomberlin (1994), pp. 431–58.

Linsky, Bernard and Zalta, Edward N. (1996). In defense of the contingently nonconcrete. *Philosophical Studies*, 84, 283–94.

Loewer, Barry (2007). Laws and natural properties. *Philosophical Topics*, 35, 313–28.

Lowe, E. J. (2010). On the individuation of powers. In *The Metaphysics of Powers: Their Grounding and Their Manifestations* (edited by Anna Marmodoro), pp. 8–26. Routledge, New York.

MacBride, Fraser (2003). Speaking with shadows: A study of neo-logicism. *British Journal for the Philosophy of Science*, 54, 103–63.

MacBride, Fraser (2005). Structuralism reconsidered. In *The Oxford Handbook of Philosophy of Mathematics and Logic*, pp. 563–89. Oxford University Press, Oxford.

Mach, Ernst (1893). *The Science of Mechanics*. Open Court, Chicago.

Markosian, Ned (1998). Simples. *Australasian Journal of Philosophy*, 76, 213–26.

Martens, Niels C. M. (2017a). Against comparativism about mass in Newtonian gravity—a case study in the metaphysics of scale. PhD thesis, University of Oxford.

Martens, Niels C. M. (2017b). Regularity comparativism about mass in Newtonian gravity. *Philosophy of Science*, 84, 1226–38.

Maudlin, Tim (1988). The essence of space-time. *PSA: Proceedings of the Biennial Meeting of the Philosophy of Science Association*, 1988, 82–91.

Maudlin, Tim (1990). Substances and space-time: What Aristotle would have said to Einstein. *Studies in History and Philosophy of Science Part A*, 21, 531–61.

Maudlin, Tim (2007). A modest proposal concerning laws, counterfactuals, and explanations. In *The Metaphysics Within Physics*, pp. 5–49. Oxford University Press, New York.

Maudlin, Tim (2014). *New Foundations for Physical Geometry: The Theory of Linear Structures*. Oxford University Press, Oxford.

McDaniel, Kris (2001). Tropes and ordinary physical objects. *Philosophical Studies*, 104, 269–90.

McDaniel, Kris (2007). Extended simples. *Philosophical Studies*, 133, 131–41.

McGee, Vann (1997). How we learn mathematical language. *Philosophical Review*, 106, 35–68.

McKenzie, Kerry (2014). Priority and particle physics: Ontic structural realism as a fundamentality thesis. *British Journal for the Philosophy of Science*, 65, 353–80.

McKenzie, Kerry (2017). Ontic structural realism. *Philosophy Compass*, 12.

McSweeney, Michaela Markham (2016). The metaphysical basis of logic. PhD thesis, Princeton University.

McSweeney, Michaela Markham (2017). An epistemic account of metaphysical equivalence. *Philosophical Perspectives*, 30, 270–93.

McSweeney, Michaela Markham (2019). Following logical realism where it leads. *Philosophical Studies*, 176, 117–39.

Melia, Joseph (1995). On what there's not. *Analysis*, 55, 223–9.

Melia, Joseph (1998). Field's programme: Some interference. *Analysis*, 58, 63–71.

Mellor, D. H. and Oliver, Alex (eds.) (1997). *Properties*. Oxford University Press, Oxford.

Miller, Elizabeth (2013). Quantum entanglement, Bohmian mechanics, and Humean supervenience. *Australasian Journal of Philosophy*, 92, 1–17.

Miller, Kristie (2005). What is metaphysical equivalence? *Philosophical Papers*, 34, 45–74.

Mumford, Stephen (2004). *Laws in Nature*. Routledge, Abingdon.

Mundy, Brent (1987). The metaphysics of quantity. *Philosophical Studies*, 51, 29–54.

Mundy, Brent (1988). Extensive measurement and ratio functions. *Synthese*, 75, 1–23.

Mundy, Brent (1989). Elementary categorial logic, predicates of variable degree, and theory of quantity. *Journal of Philosophical Logic*, 18, 115–40.

Mundy, Brent (1992). Space-time and isomorphism. *PSA: Proceedings of the Biennial Meeting of the Philosophy of Science Association*, 1992, 515–27.

Nagel, Ernest (1961). *The Structure of Science: Problems in the Logic of Scientific Explanation*. Harcourt, Brace & World, New York.

Ney, Alyssa (2014). Review of Steven French, *The Structure of the World*. *Notre Dame Philosophical Reviews*, 2014.

Nolan, Daniel (1997). Quantitative parsimony. *British Journal for the Philosophy of Science*, 48, 329–43.

Nolan, Daniel (2011). The extent of metaphysical necessity. *Philosophical Perspectives*, 25, 313–39.

North, Jill (2009). The 'structure' of physics: A case study. *Journal of Philosophy*, 106, 57–88.

North, Jill (2018). A new approach to the relational-substantival debate. In *Oxford Studies in Metaphysics* (edited by Karen Bennett and Dean W. Zimmerman), volume 11, pp. 3–43. Oxford University Press, Oxford.

Parsons, Josh (2007). Theories of location. In *Oxford Studies in Metaphysics* (edited by Dean W. Zimmerman), volume 3, pp. 201–32. Oxford University Press, Oxford.

Paul, L. A. (2002). Logical parts. *Noûs*, 36, 578–96.

Paul, L. A. (2012a). Building the world from its fundamental constituents. *Philosophical Studies*, 158, 221–56.

Paul, L. A. (2012b). Metaphysics as modeling: The handmaiden's tale. *Philosophical Studies*, 160, 1–29.

Paul, L. A. (2017). A one category ontology. In *Being, Freedom, and Method: Themes From the Philosophy of Peter van Inwagen* (edited by John Keller), pp. 32–61. Oxford University Press, Oxford.

Plantinga, Alvin (1976). Actualism and possible worlds. *Theoria*, 42, 139–60.

Pooley, Oliver (2006). Points, particles, and structural realism. In *The Structural Foundations of Quantum Gravity* (edited by Dean Rickles and Steven French and Juha Saatsi), pp. 83–120. Oxford University Press, Oxford.

Psillos, Stathis (2001). Is structural realism possible? *Proceedings of the Philosophy of Science Association*, 2001, 13–24.

Psillos, Stathis (2006). The structure, the whole structure, and nothing but the structure? *Philosophy of Science*, 73, 560–70.

Putnam, Hilary (1981). *Reason, Truth and History*. Cambridge University Press, Cambridge.

Quine, W. V. (1948). On what there is. *Review of Metaphysics*, 2, 21–38.

Quine, W. V. (1950). Identity, ostension, and hypostasis. *Journal of Philosophy*, 47, 621–33.

Quine, W. V. (1960a). Variables explained away. *Proceedings of the American Philosophy Society*, 104, 343–7.

Quine, W. V. (1960b). *Word and Object*. MIT Press, Cambridge, MA.

Quine, W. V. (1961). Reply to Professor Marcus. *Synthese*, 13, 323–30.

Quine, W. V. (1964). Review of Geach, *Reference and Generality*. *Philosophical Review*, 73, 100–4.

Quine, W. V. (1970). *Philosophy of Logic*. Harvard University Press, Cambridge, MA. Second edition, 1986.

Quine, W. V. (1975). On empirically equivalent systems of the world. *Erkenntnis*, 9, 313–28.

Quine, W. V. (1976). Grades of discriminability. *Journal of Philosophy*, 73, 113–16.

Rayo, Agustín (2013). *The Construction of Logical Space*. Oxford University Press, Oxford.

Reichenbach, Hans (1958). *The Philosophy of Space and Time*. Dover, New York.

Resnik, Michael D. (1981). Mathematics as a science of patterns: Ontology and reference. *Noûs*, 15, 529–50.

Resnik, Michael D. (1997). *Mathematics as a Science of Patterns*. Oxford University Press, New York.

Robinson, Howard (1982). *Matter and Sense: A Critique of Contemporary Materialism*. Cambridge University Press, Cambridge.

Rodriguez-Pereyra, Gonzalo (2002). *Resemblance Nominalism: A Solution to the Problem of Universals*. Clarendon Press, Oxford.

Rosen, Gideon (2010). Metaphysical dependence: Grounding and reduction. In *Modality: Metaphysics, Logic, and Epistemology* (edited by Bob Hale and Aviv Hoffmann), pp. 109–36. Oxford University Press, Oxford.

Rosen, Gideon (2017). What is a moral law? *Oxford Studies in Metaethics*, 12, 135–59.

Russell, Bertrand (1905). On denoting. *Mind*, 14, 479–93.

Russell, Bertrand (1912). *The Problems of Philosophy*. Williams and Norgate, London. Paperback edition by Oxford University Press, 1959.

Russell, Bertrand (1927). *The Analysis of Matter*. Routledge, London.

Russell, Bertrand (1940). *An Inquiry into Meaning and Truth*. Allen and Unwin, London.

Russell, Jeffrey Sanford (2015). Possible worlds and the objective world. *Philosophy and Phenomenological Research*, 90, 389–422.

Russell, Jeffrey Sanford (2017). Qualitative grounds. *Philosophical Perspectives*, 30, 309–48.

Russell, Jeffrey Sanford (2018). Quality and quantifiers. *Australasian Journal of Philosophy*, 96, 562–77.

Saatsi, Juha (2010). Whence ontological structural realism? In *EPSA Epistemology and Methodology of Science* (edited by Mauricio Suárez, Mauro Dorato, and Miklós Rédei), pp. 255–65. Springer, Dordrecht.

Saucedo, Raul (2011). Parthood and location. In *Oxford Studies in Metaphysics: Volume 5* (edited by Dean Zimmerman and Karen Bennett). Oxford University Press.

Saunders, Simon (2003). Indiscernibles, general covariance, and other symmetries. In *Revisiting the Foundations of Relativistic Physics. Festschrift in Honour of John Stachel* (edited by Abhay Ashtekar, Jürgen Renn, Don Howard, Abner Shimony, and S. Sarkar), pp. 151–73. Kluwer, Dordrecht.

Saunders, Simon (2016). On the emergence of individuals in physics. In *Individuals Across The Sciences* (edited by Thomas Pradeu and Alexandre Guay). Oxford University Press, Oxford.

Saunders, Simon and McKenzie, Kerry (2015). Structure and logic. In *Physical Theory: Method and Interpretation* (edited by Lawrence Sklar), pp. 127–62. Oxford University Press, Oxford.

Schaffer, Jonathan (2005). Quiddistic knowledge. *Philosophical Studies*, 123, 1–32.

Schaffer, Jonathan (2009). On what grounds what. In Chalmers et al. (2009), pp. 347–83.

Schaffer, Jonathan (2010). Monism: The priority of the whole. *Philosophical Review*, 119, 31–76.

Schaffer, Jonathan (2016). Grounding in the image of causation. *Philosophical Studies*, 173, 49–100.

Schaffer, Jonathan (2017a). The ground between the gaps. *Philosophers' Imprint*, 17, 1–26.

Schaffer, Jonathan (2017b). Laws for metaphysical explanation. *Philosophical Issues*, 27, 302–21.

Shapiro, Stewart (1997). *Philosophy of Mathematics: Structure and Ontology*. Oxford University Press, New York.

Shoemaker, Sydney (1980). Causality and properties. In *Time and Cause* (edited by Peter van Inwagen), pp. 228–54. D. Reidel, Dordrecht.

Shumener, Erica (2020). Explaining identity and distinctness. *Philosophical Studies*, 177, 2073–96.

Sidelle, Alan (1989). *Necessity, Essence and Individuation*. Cornell University Press, Ithaca, NY.

Sider, Theodore (1996). Naturalness and arbitrariness. *Philosophical Studies*, 81, 283–301.

Sider, Theodore (2001). *Four-Dimensionalism*. Clarendon Press, Oxford.

Sider, Theodore (2006a). Bare particulars. *Philosophical Perspectives*, 20, 387–97.

Sider, Theodore (2006b). Beyond the Humphrey objection. Available at <http://tedsider.org/papers/counterpart_theory.pdf>.

Sider, Theodore (2007). Neo-Fregeanism and quantifier variance. *Aristotelian Society, Supplementary Volume*, 81, 201–32.

Sider, Theodore (2008a). Monism and statespace structure. In *Being: Developments in Contemporary Metaphysics* (edited by Robin Le Poidevin), pp. 129–50. Cambridge University Press, Cambridge.

Sider, Theodore (2008b). Yet another paper on the supervenience argument against coincident entities. *Philosophy and Phenomenological Research*, 77, 613–24.

Sider, Theodore (2009). Ontological realism. In Chalmers et al. (2009), pp. 384–423.

Sider, Theodore (2011). *Writing the Book of the World*. Clarendon Press, Oxford.

Sider, Theodore (2013). Against parthood. In Bennett and Zimmerman (2013), pp. 237–93.

Sider, Theodore (2020). Ground grounded. *Philosophical Studies*, 177, 747–67.

Sklar, Lawrence (1982). Saving the noumena. *Philosophical Topics*, 13, 89–110.

Skow, Bradford (2008). Haecceitism, anti-haecceitism and possible worlds. *Philosophical Quarterly*, 58, 98–107.

Stachel, John (1993). The meaning of general covariance. In *Philosophical Problems of the Internal and External Worlds: Essays on the Philosophy of Adolph Grünbaum* (edited by John Earman, A. Janis, and G. Massey), pp. 129–60. University of Pittsburgh Press.

Stachel, John (2002). 'The relations between things' versus 'The things between relations': The deeper meaning of the hole argument. In *Reading Natural Philosophy: Essays in the History and Philosophy of Science and Mathematics* (edited by David B. Malament), pp. 231–66. Open Court, Chicago.

Stalnaker, Robert (1977). Complex predicates. *The Monist*, 60, 327–39.

Stalnaker, Robert (1984). *Inquiry*. MIT Press, Cambridge, MA.

Stalnaker, Robert (2012). *Mere Possibilities: Metaphysical Foundations of Modal Semantics*. Princeton University Press, Princeton, NJ.

Steward, Stephen (2016). On what consists in what. PhD thesis, Syracuse University, NY.

Strawson, Peter F. (1959). *Individuals: An Essay in Descriptive Metaphysics*. Routledge, London.

Sud, Rohan (2018). Vague naturalness as ersatz metaphysical vagueness. *Oxford Studies in Metaphysics*, 11, 243–77.

Swoyer, Chris (1982). The nature of natural laws. *Australasian Journal of Philosophy*, 60, 203–23.

Tarski, Alfred and Givant, Steven (1999). Tarski's system of geometry. *Bulletin of Symbolic Logic*, 5, 175–214.

Teitel, Trevor (2019). Holes in spacetime: Some neglected essentials. *Journal of Philosophy*, 96, 353–89.

Teller, Paul (1987). Space-time as a physical quantity. In *Kelvin's Baltimore Lectures and Modern Theoretical Physics* (edited by P. Achinstein and R. Kagon), pp. 425–48. MIT Press, Cambridge, MA.

Thomasson, Amie L. (2007). *Ordinary Objects*. Oxford University Press, New York.

Thomasson, Amie L. (2015). *Ontology Made Easy*. Oxford University Press, New York.

Tomberlin, James (ed.) (1994). *Philosophical Perspectives 8: Logic and Language*. Ridgeview, Atascadero, CA.

Torza, Alessandro (2017). Ideology in a desert landscape. *Philosophical Issues*, 27, 383–406.

Turner, Jason (2011). Ontological nihilism. In *Oxford Studies in Metaphysics* (edited by Karen Bennett and Dean W. Zimmerman), volume 6, pp. 3–54. Oxford University Press, Oxford.

Turner, Jason (2016a). Can we do without fundamental individuals? No. In Barnes (2016), pp. 24–34.

Turner, Jason (2016b). *The Facts in Logical Space: A Tractarian Ontology*. Oxford. Oxford University Press.

Turner, Jason (2021). On doing without ontology: Feature-placing on a global scale. In *The Question of Ontology* (edited by Javier Cumpa). Oxford University Press. Forthcoming.

Van Cleve, James (1985). Three versions of the bundle theory. *Philosophical Studies*, 47, 95–107.

van Inwagen, Peter (1990). *Material Beings*. Cornell University Press, Ithaca, NY.

Wallace, David (2012). *The Emergent Multiverse*. Oxford University Press, Oxford.

Warren, Jared (2016). Sider on the epistemology of structure. *Philosophical Studies*, 173, 2417–35.

Weatherall, James Owen (2016). Are Newtonian gravitation and geometrized Newtonian gravitation theoretically equivalent? *Erkenntnis*, 81, 1073–91.

Whittle, Ann (2008). A functionalist theory of properties. *Philosophy and Phenomenological Research*, 77, 59–82.

Williams, Donald C. (1952). On the elements of being. *Review of Metaphysics*, 7, 3–18, 171–92. Part I reprinted in Mellor and Oliver (1997, 112–24).

Williamson, Timothy (1985). Converse relations. *Philosophical Review*, 94, 249–62.

Williamson, Timothy (1998). Bare possibilia. *Erkenntnis*, 48, 257–73.

Williamson, Timothy (2000). *Knowledge and its Limits*. Oxford University Press, Oxford.

Williamson, Timothy (2002). Necessary existents. In *Logic, Thought and Language* (edited by A. O'Hear), pp. 233–51. Cambridge University Press, Cambridge.

Williamson, Timothy (2013). *Modal Logic as Metaphysics*. Oxford University Press, Oxford.

Wilsch, Tobias (2015). The nomological account of ground. *Philosophical Studies*, 172, 3293–312.

Wilson, Jessica M. (2014). No work for a theory of grounding. *Inquiry*, 57, 535–79.

Wilson, Jessica M. (2018). Grounding-based formulations of physicalism. *Topoi*, 37, 495–512.

Wolff, J. E. (2020). *The Metaphysics of Quantities*. Oxford University Press, Oxford.

Worrall, John (1989). Structural realism: The best of both worlds? *Dialectica*, 43, 99–124.

Wright, Crispin (1983). *Frege's Conception of Numbers as Objects*. Aberdeen University Press, Aberdeen.

Index

Printed and bound by CPI Group (UK) Ltd, Croydon, CR0 4YY